Limited Edition

For Phil

ANGEL AND DAMIEN

By

Brianna Io Rivera

Printed in the United States of America

First Edition: 2024

Table of Contents

PART ONE

The bitch will pay

Damien scowled at the crescent moon hanging above him. He crushed the remains of the rodent he had just drained and hurled it against a tree. His wretched state of filth feeding from filth was not the most difficult part of his escape. It had taken him five of the six decades of his imprisonment just to crack the runes warding his prison.

The fact they couldn't destroy him completely provided little comfort as he lay imprisoned and starving for sixty-three years. The only thing which kept him from madness was his fantasies of how he would destroy the one responsible. Regardless of who her family was.

His eyes swirled in shades of red and black, brimming with ire. He wiped his face, brushed off his shoulders, and turned south. A small smile tugged at the corner of his mouth as he took his first step toward exacting his revenge.

~

"Asshole!" I shouted at the open manhole into which the slime demon had just slithered.

I shook my arms, flicking bits of goo onto the pavement.

"Always with the gross," I muttered to myself, wiping my hands on my pants. I regarded the hole in the street and sighed.

When I pulled the Devil card in my Tarot reading this morning, I did not expect it to be so literal. If that was how the cards wanted to play it today, I wondered whether drawing the Tower meant I was going to be buried in rubble from the sewer collapsing on me. The prospect almost seemed better than worrying about whether I would once again need to rebuild what I had. Seeing as I had pulled the Six of Swords as the third card in the three-card reading, I couldn't help but worry I'd soon need to relocate to another city. I blew a stray lock of blonde hair out of my eyes. This was not the time. I would soon see what the reading meant, and focus was what I needed right now.

I leaped into the manhole, landing with a splash in the sewer. Looking around and desperately trying to ignore the fact that I was standing in the intestines of the city, I spotted the trail of slime heading north.

"You okay down there, Pepper?" Kaden, my best friend and partner, called down from above. My name was Angelica, but he'd called me Pepper since we were eight.

One little cooking disaster and you're "Pepper" for the rest of your life.

"Just ducky, K. Are you coming down, or are you going to stay up there like a princess?"

"Feel free to call me Her Royal Highness. I have this weird thing about E. coli."

"Ugh. Fine, you asshat."

I shook my hands again, this time to focus on my battle magic. I had originally planned to bring the slime demon in, but the prick had ruined the game by ruining my shirt. Since his bounty paid "dead or alive," I wasn't going to lose sleep over killing this one.

I tracked the creature's path relatively easily. Slime demons can be cagey, but they are not known for their intelligence. It took me less than twenty minutes to find him about fifty yards in, tucked into a recess in the ceiling. Not wanting to literally bring the city down on my head, I had to go for a more subtle approach than just blasting the fucker. I reached out a hand and began to squeeze.

The demon's eyes flew open and began to widen comically. His mouth dropped open, and viscous fluid hit the ground in wet blobs. I jerked my hand backward, and he popped from the ceiling with a sucking noise, falling to my feet.

"Hi there," I greeted him with a facetious grin, slightly letting up on the squeeze. Sometimes a girl could get valuable information out of a mark. And information can be the real payday.

"Anything you'd like to confess before I pop you like a zit?"

And then it happened.

He smiled and said the words I'd been dreading since that night.

"Damien's free, you cunt."

I felt the blood drain from my face.

"And he's coming for you, little girl."

I ended the demon with a little more force than necessary, but the bile which rose to my throat had nothing to do with his innards hitting my face.

~

DAMIEN

Damien sat on the rooftop, watching the people in the streets below. He frowned at his clothing. He wanted his first feed to be just right, so he had snapped the neck of a man in the alley for his clothes. The cheap material chafed his body and his pride.

More reason to punish the girl.

A decade of feeding on earth vermin to gain enough strength to dig himself out of the hole in which the wretched girl had entombed him had cultivated his vengeance into a fever. An obsession.

But he couldn't take her yet. He needed to recover his strength. And then he would tear her throat out.

Giggling caught Damien's attention and shook him from his fantasy. He reached between his legs and felt his cock straining against his stolen pants. He smiled. Just the thought of destroying that girl had made him hard. He glanced down at the giggling girls, and his smile widened. His first feed was going to be even more fun than he'd planned.

~

One Bourbon, One Scotch, One...Nah, I'll Just Go Home

I sat with my elbows on the lacquered bar top, staring into my drink. The news that Damien was awake and undoubtedly on a very specific hunt was terrifying to the point of debilitation.

"Maybe it's not even true, Pepper," Kaden said, sitting next to me and leaning on one arm to face me.

"You know that's not the case, K," I replied, continuing to stare into my drink as if it would yield a solution.

"Demons lie, Pep! He was just trying to get into your head!"

"Yes. But the cards don't. I pulled the Devil, the Tower, and the Six of Swords. Obsession, Ruin, and Travel. The message is loud and clear."

I turned to face my friend, and my heart clenched at the hope in his face. I shook my head and redirected my attention to the bar, sighing.

"I want to believe the demon was lying. I really do. But I don't. And even if I did, it would be stupid and negligent to blow off something like this, and you know it, K."

Kaden considered that for a moment before nodding.

"You're right. So, what's next?"

"Well, that's just it. I don't know. Everyone who helped us before is either dead or retired. Or… *unwilling* to help." I scrubbed my face with my hands. We know where he was last, obviously, but to go after him without any backup is another item on the list of stupid things I'd rather not do."

I grabbed my drink and tossed it back. A headache started behind my eyes, rendering the idea of another drink unsavory, so I reached for my wallet to settle my tab.

"Well, we have another job tomorrow, so maybe the distraction will help," Kaden said, also pulling out his wallet.

We had planned to walk back to my place together and sketch out tomorrow's job.

"Maybe," I said quietly to myself.

"Hey," Kaden said, dipping his face down to look me in the eyes. I looked up at him.

"It's going to be okay, Pepper. We're going to figure something out." He put a hand on my shoulder. "We always do."

I nodded, but the knot of dread which had tied itself into my stomach when the slime demon mentioned Damien refused to untie.

"Okay," I replied, nodding again, and rose from the barstool.

Draco Winter

Draco watched her over his glass and blinked his hazel eyes languidly. The Fae's nerves had her prattling, and he tried to keep his irritation in check. Typically, he would not have the patience for her company, but she was the only supplier of phoenix blood he knew of, and he was nearly out.

Her nervousness was wreaking havoc on her control, and every so often, he would catch a sliver of her true form beneath her glamour: a hint of sharpened canines, white-blonde hair, and silver eyes. She was one of the few Fae who would sell outside of the Courts. Expensive and insufferable but discreet and reliable, so he did his best to hold his tongue.

But even he had his limits.

"Morgaina."

She started slightly and snapped her jaws shut mid-sentence. His eyes fluttered closed briefly as he enjoyed the silence.

"I would like to get on with my day if you are quite finished."

She nodded mutely, flushing slightly.

He'd met Morgaina at a bonding ceremony many years back. After a couple of cups of nectar, she'd thought herself brave and had propositioned him, assuming her Fae wiles were irresistible. He'd summarily dismissed her. She had darted off in mortified tears, and he'd watched her stumble away, finding her entertaining for the first time that night. She evidently never got over it.

There was only one for him. Regrettably, she couldn't stand the sight of him.

He was, however, a patient man. An impulsive mage lived a short life, and he had lived many lifetimes over. And *she* was worth the wait. Just the thought of her made his blood warm. Before he lost himself to his daydreams, he drew his attention back to the business at hand.

He casually threw a rune to envelop them in an insulating spell before sliding her payment across the table, just out of reach. He kept his hand over the envelope, indicating she produced her wares. She did so quickly, nearly knocking his drink to the floor in her haste. He glanced at the glass before dragging his eyes back to her face.

"I'm sorry, Draco," she squeaked, trembling slightly.

Tendrils of darkness whispered from his fingertips before he coaxed them back. He didn't have time for this.

"We're done here."

He retrieved his phoenix blood and strode out without sparing her another glance.

He tucked the package in his coat pocket as he stepped back into the bustling womb of Los Angeles. After the soft lighting in the Fae bar, the afternoon sunlight stung his eyes. Most members of the supernatural community avoided human cities with their noise and filth, but Draco loved them. He'd been to most during his long life and found he reveled in the anonymity and the palpable energy hordes of people provided. He'd had his fill of celebrity and had eschewed it centuries ago. When your name was on too many lips, trouble was sure to follow.

He ducked into an alleyway, his footsteps echoing in the gloom. He stopped to stroke the downy head of a small, silver tabby who had trotted up to him.

“Hello, my friend.”

Draco had never had a familiar before and had given up on the idea when Mischief found her way into his life. He soon came to adore his unusual friend. As a tiny kitten, she’d contracted a virus which had damaged her motor skills, meaning she walked like a drunk and would fall over without warning. It had also unlocked something special, allowing her to communicate with a witch or mage.

Hi, friend. You were gone a long time.

“Amorous, brainless, fae.”

The little gray head nodded knowingly.

But the only contact with phoenix blood.

“Regrettably.”

So, worth it, no?

Draco sighed his assent.

Her tail twitched, and her back end wobbled before she turned and headed back into his shop.

Cats

DAMIEN

Damien licked his fingers. He had drained the first girl immediately after satisfying his lust with the two of them but not before compelling her to bind the other to the heavy oak chair. He closed his eyes in bliss and took in the familiar scents with a deep breath.

It was good to be home.

He had taken them back to his villa and, wanting to wash away the dirt and filth from his body in addition to enjoying some carnal pleasures, began the evening in the shower. Two birds, one stone, he figured. The first "bird" now sat dead on the chaise, unseeing eyes trained on the frescoed ceiling. The second whimpered behind him. He rolled his eyes.

He wrinkled his nose when his gaze landed on the pile of clothing he'd taken from the man in the alley. He lifted his hand and snapped his fingers, igniting the clothing into a ball of blue flame. He watched with satisfaction as the cheap garments burned to ash.

The sharp scent of urine stung his nose, causing him to turn his disdain in the direction of the prostrated girl. She was trembling now, whether in fear or cold. Damien was unsure, but either was satisfactory. Her nipples strained against her skin, and the flesh around them was turgid and pimpled. He brought his gaze slowly back to her face and smiled. The trembling turned to violent shaking, and she began blubbering. He raised one eyebrow.

Fear, then.

Her eyes widened as he began to saunter over to her, and fat tears splashed onto her chest. He drew closer, and she began to shake her head wildly.

“No, no, no, please…” She pushed against the back of the chair in an instinctive attempt to get some distance from him.

“Shhh...shh, shh, shh…” Damien pressed his finger on her lips before dragging it down her chin, painting a line to her neck with her friend’s blood. He tipped his head to the side to consider the hollow of her throat before settling his hand around her neck. He tapped his thumb on her collarbone before dropping his hand and straightening. He turned around and headed toward his bedroom, wanting to be dressed in fine clothing once more. After a few steps, he stopped and said, “Sit tight,” over his shoulder before chuckling and shaking his head at his own little joke.

Very, very good to be home, indeed.

Pep Talk

"Should be a routine gig. In and out, no mess."

"That's what she said." My automatic response was half-hearted at best.

Kaden regarded me silently before he slowly laid his pen on the sketches of the warehouse we were to investigate tomorrow.

"Pepper," he said.

"Yes." I stared at my hands folded in my lap. I knew he wanted me to look at him so he could use those big, brown doe-eyes and hit me with his best friend mojo.

"Angelica," he said softly.

That caused me to look up at him. He rarely used my actual name.

"Kaden?"

"How is freaking out and stressing yourself to the point of peptic ulcers going to help against Damien?" I gave him a nonplussed look. "He's an amoral monster, yes. He's a horrifying, violent psychopath. And immortal."

"You going somewhere with this, K?"

"Yes." He leaned his forearms on his thighs. "You've beaten him once before. That means not only do we know it can be done, but I guarantee you've got some serious rental space in his head."

I pursed my lips. He had a point.

"That actually does make me feel a little better. But how is psychological warfare enough to substitute for actual warriors?" I shook my head. "We need more than just the two of us, K."

"I know. And I think I may have an idea about that."

Damien's Phone Call

Damien held the gilded bowl beneath the girl's chin, catching her blood as it poured from her slit throat. Her whining and crying had grated on his last nerve, so he silenced her. Now she sat with her life draining into the bowl, her mouth opening and closing like a fish. He savagely bit into his wrist, jerking his head to the side as he felt a vein between his teeth. Damien savored the pop it made, the vessel's jagged edges quivering and pulsing with each beat of his heart.

He watched the small crimson river wend its way through his fingers. The pain in his wrist receded to a dull throbbing, and the wound itched as it knitted itself back together. He slid his tongue out to catch an errant rivulet tickling his lip before guiding the stream with his fingertips to mingle with the girl's blood. He waved his hand over the bowl, drawing intricate runes in the air while Latin rumbled from his chest. After regarding his reflection for a moment, he snapped the spell closed, and another face replaced his on the surface of the liquid.

"Exaudi me Domine mi. Reversus sum ad vos."

Hear me, my Lord. I have returned to you.

He was certain the Master was unaware of his return, so Damien wanted to ensure he established the connection and proved his continued loyalty. He shuddered at the thought of how furious the Master might be after his silence of over sixty years, especially with his vile brother whispering doubts in His ear during his absence. Damien's brother, Malachi, was a right prick.

Power began to pinch his skin, sending small electric shocks through his body. He shuddered as the voice slithered down his spine before booming in his head.

COWARD.

The word echoed painfully. Damien's brow furrowed. His brother was never one to disappoint. The cocksucker.

No, my Lord. I was captured and imprisoned, unable to reach out to you.

SOMEONE DEFEATED YOU. YOU ARE WEAK.

He hung his head and inched closer to the bowl.

I am weak no longer. Hate and vengeance fuel me with righteous fury. She will suffer.

SHE??

Dread filled him, and he winced. He had intended to leave that detail out.

She had an army, my Lord. Armed with magic—

YOU BRING LEGION EXCUSES USELESS AS YOU.

My Lord...

IS THIS WHY YOU CALLED US? TO PEDDLE EXCUSES AND SOLICIT DISAPPOINTMENT? YOU ARE NEEDED NO LONGER.

Wait! How can I redeem myself? What do I need to do?

BRING US THE FEMALE.

The power was abruptly sucked out of the room. Blood sloshed over the side of the bowl as Damien pushed it away. Shame and frustration filled him.

“Etiam, Pater,” he said through gritted teeth.

Yes, Father.

Wren Vander

"Okay, I'll bite. What's your idea?"

"You're not gonna like it."

"Dude," I said, giving him an incredulous look, "Back, meet Wall. Wall, Back. I see you've already met Rock, Hard Place, and Totally Fucked."

"Cute."

I grunted at him.

"Well," Kaden cleared his throat, "when, uh, was the last time you gave the family a call?"

I swung my head around and stared at him.

"You're right. I don't like it."

"Just hear me out, Pepper—"

"No," I cut him off.

"But it's just—"

"No."

"Pepper, damn it, let me talk!"

That shut me up. Kaden rarely lost his cool and almost never with me. In fact, in the centuries we'd been friends, I could probably count on one hand the number of times he'd even raised his voice at me. Even he looked a little shocked.

"I'm sorry, Pepper."

"Don't be. I was being an ass. Please tell me your idea. I promise I will interrupt as little as possible."

That earned me a smile.

"The family."

I nodded.

"You know I don't suggest this lightly. But can you think of anything better? Personally, I can't think of a single viable alternative, and I'm scared Pep. Really, really scared."

I regarded him silently before letting out a resigned sigh.

"Me too, K," I said, slumping. "This is so much bigger than my relationship with my horrible family. It's a resource I would be stupid to ignore because I'm angry and...other things." I flopped back against the couch cushions and closed my eyes. "But you know, for there to be any chance of getting them to help, you should probably be the one to reach out." I opened one eye to look at him. "They actually like you."

He answered by taking my hand and giving it a light squeeze.

Every family has issues. I was not so self-absorbed as to think myself unique in that respect. Mine, however, was kind of like the Addams Family without the love and devotion. My father, Rache Angelicus, was a very old, very scary vampire from a very long time ago. His name translates to "revenge" in German, though I'm not sure that was the name he was born with. I never thought to ask.

My mother, Wren, was even scarier. As she tells it, she began her life in the 17th century, in a small village somewhere in the middle of Nowhere, Europe. As she matured, her parents discovered she was a witch. Of course, in those times, such a thing was a death sentence, and her parents did not disappoint. Terrified of her and the powers they did not understand, they immediately turned to the new religion sweeping the nation, which promptly tried to burn her alive.

On the day of her "trial," she was deemed guilty of "engaging in mystical dark magics and lewd sexual acts with the Devil" and was sentenced to "cleansing" by way of fire. It did not go as planned. Instead of burning this waif of a fifteen-year-old girl on a pyre erected in the village square, it ended up being the last day on earth for every one of the villagers.

They marched her, bound and gagged, to the center of the village and up a set of crudely fashioned steps to a pyre in the final stages of construction. Leaving her tied, they doubled the restraints by wrapping rope around her midsection and feet. A third rope was strung around her neck and secured to the large log behind her. She made no move to stop them.

They removed the gag from her mouth and asked her to confess and save her immortal soul. Instead of pleading her innocence or begging for release, she looked her parents in the eye individually and gave them a slow smile. She maintained eye contact with each, sliding her gaze back and forth as the Village Elder and self-appointed spiritual leader rambled about the 'Almighty,' 'Righteousness,' and 'Eternal

Salvation' while the sycophantic rabble lit the kindling with perverse excitement.

The ropes were tight, but she was unafraid. She had an intimate relationship with the power inherent in fire and its destructive elegance. Instead of harnessing the flames tickling her toes, she looked into the savagely expectant faces of the crowd. They had bound her arms behind her at the wrists, leaving her hands plenty of mobility to cast. First, she disintegrated the ropes binding her wrists without letting the crowd see, allowing the drama and the flames to build.

Then, she began to cast runes, drawing storm clouds crackling with electricity and pregnant with rain, dark enough to blot out the sun. Once the villagers started looking uneasy, she glanced at her parents one more time before casting a lightning strike which exploded the massive oak celebrated as the centerpiece of the village square. She parted her lips and whispered while she cast, acting as a ventriloquist and a booming voice shook the village. The villagers scattered in panic, to my mother's delighted satisfaction.

"WHO DARES DESTROY THIS INNOCENCE? YOU WILL ALL SUFFER MY WRATH! I AM YOUR ALMIGHTY GOD, AND I HEREBY DAMN EVERY

SOUL HERE TO ETERNAL TORMENT IN THE FLAMES OF HELL!"

Pandemonium ensued. Screams rent the air as my mother cast and cast, stifling her laughter while sending lightning strikes and torrential rains. Once every single villager was dead, save her parents, she cast a rictus spell to secure them in place. Only then did she dissolve the rest of the rope binding her. She cast a rune and absorbed the remaining fire.

Her parents stood before her, terrified statues. The only thing she allowed them to move for now was their eyes. The eyes willing to watch their precious girl burn in the name of ignorance and superstition. She hopped down from the pyre and strode casually to stand in front of them.

She peered into their faces.

"Lingua," she said with a bored expression, releasing their ability to speak. Her mother began blubbering, and her father began praying, of all things. She shook her head.

"You jest. This is a farce you've cooked up to mock me in your last moments. I appreciate the irony, but I would imagine you'd want to spend your last breaths on something worthwhile." She shrugged. "To each his own." With a rough gesture, she cast a spell, and her father's neck snapped with such violence that his head turned nearly one hundred eighty degrees.

"Mother." She turned to take in her mother's horrified, heartbroken face and wide, weeping eyes. "You could have stopped this, you know," she began softly. "This," she gestured wide to include the carnage around them and the wreckage of the demolished village, "is all your fault. It could have been avoided if you hadn't been…" Her expression, initially void of emotion, became hateful.

"Such a bloody, selfish simp."

She threw a final rune, and the ground beneath her mother's feet caught fire. As the flames began to creep, she turned away and left, the screams her last memory of childhood. At this point in the story, my mother would always don the expression of a misunderstood and heartbroken child. She would always end the story with, "It was difficult to leave her that way. Her screams still haunt me."

Any guests in our house, enthralled by her story, would nod with sympathetic smiles and pat her on the shoulder as she dabbed at her eyes. They never caught the real moral of the story. All which unbelievable power at fifteen.

With a simple rune cast, my mother could have stopped the screaming.

Rache Angelicus

My father was different. Not only was he less forthcoming with his history, but he was cold in body and spirit. What I do know is spotty but unpleasant enough that I became unwilling, from a fairly young age, to dig too deeply.

As I've already stated, his name is Rache, but I believe it to be a name he adopted later in his life. He was legendary for his cruelty and violence. One story I did know came from later in his life when the Christian Church had established a stronghold in Western Europe.

It was a lovely autumn afternoon, and the trees were dappled with sunlight. Contrary to popular belief, vampires, especially old ones like my father, are not going to combust from stepping outside for an afternoon constitutional on a bright summer day. Even the newly born are immune. The myth began due to the fact which younglings avoid the sun until they adjust to their enhanced senses.

So, Father was strolling along a cobbled path, enjoying the trees and their fireworks display. Shuffling through the fallen leaves, he caught the scent of a gathering of people tucked into a charming church with stained glass and sweeping slate steps leading to the front door.

He heard music, and it tickled his senses, so he followed the sound to the large doors of the church. Opening the doors wide, he was greeted with the sight of his cattle singing, already on their knees. Whistling a merry tune, he turned around, closed the doors, lifted a bronze candelabra, slid it gently through the door handles, and twisted the metal into a knot before turning back around to his now silent buffet with a big smile. He kept them there for three weeks. He even brought them food and water when the church ran dry. That was when he developed his infamous taste for nuns.

My father was already centuries old when he heard the word of my mother's escapades and was intrigued. He tracked her doggedly across the known world until, after six years, he hunted her down. He swears to this day he no longer needs to breathe because she still holds it after stealing it from him when he saw her for the first time. Short version: soulmates, whirlwind romance, a whole lot of destruction, mayhem, and bloodletting, and here comes me. An anomaly.

Vampires aren't supposed to be able to beget children. Something to do with their technically being dead. But apparently, my mother's magic ignited something in dear old dad's nethers, which sparked his swimmers to life long enough to conceive me. I've tried not to overthink it.

The aspect of their coupling for which I was grateful was that I inherited all of mother's magic with none of father's penchant for Bloody Marys. I did get some of the ridiculous strength and an even longer lifespan than the average witch of approximately a millennium.

Shit's gone way south if I was even considering the family for help. I didn't even go to them when I clashed with Damien the first time. But this was different. I lost my werewolf connection when the entire pack, whose help I enlisted, was wiped out, including my boyfriend at the time, Colton. Two wolves survived. Needless to say, they were not jumping out of their fur to back me.

My family wasn't the best option; it was the only option.

Family Matters

Kaden and I drove to my family's estate in relative silence. I watched the trees and farmlands whiz by my window, fiddling with my tarot deck to keep from scowling.

"Do your jaws hurt?" Kaden glanced at me before returning his attention to the road.

I sighed. "Meaning?"

"You're clenching your teeth so hard, I'm surprised one hasn't cracked."

"I don't think I need to tell you why."

"No, you don't." He paused. "This is the right move, Pepper."

I squeezed my eyes shut. I knew this was necessary. We needed reinforcements. Still, dread weighed heavily on my chest. Whenever I dwelled too much on our destination, my stomach flipped. I'd already made Kaden pull over twice out of fear of having an Exorcist moment.

We continued in silence for a while longer. As the landscape became increasingly familiar, my heart began hammering against my ribcage.

"Turn around," I said in a small voice.

"What?"

"Turn around. I can't do this."

"Pepper…"

"No. This is a terrible idea." I covered my face with my hands. Tears threatened.

"Pepper, it's going to be okay."

"How? How is it possibly going to be okay?"

"Well... I'm not sure, but it doesn't really matter now."

"Huh? Oh."

Standing in the center of the road was my father. He stood with his arms crossed and his legs spread. His smile was like that of a cat who ate the canary. Actually, it's more like the shark who ate the canary. Even the little wave he gave us bespoke of a predator. We rolled to a stop in front of him and lowered the windows.

"I brought both drivers so you can ride with me in the limo. Michael will follow us in your car, Kaden."

"And hello to you too, father." Suddenly, I was exhausted. Being around my father had that effect. It felt as if I were trying to sleep to escape a nightmare.

"Hello, Angelica." He made a show of checking his watch. "We haven't much time. Dinner is at eight this evening."

The ride home was both interminable and too short. My father never was one for small talk, and that was all we had. Soon, the wrought-iron gates encircling the property came into view. Gravel popped as we drew up to the sweeping front stairs. We opened the limo doors and handed Michael our sparse luggage. I'd intentionally packed lighter than usual as if hoping to shorten the visit by packing less. Besides, there was no shortage of baggage where we were going.

A Hunting, We Will Go

After disposing of the bodies of the two young girls, Damien took another shower, this time alone. He towelled off and swiped a hand across the mirror before readying his shaving kit. He threw away the rusted blade and slid a fresh one into his razor. The last time he'd shaved had been the morning of the night *she* had trapped him for over sixty years. Razor in hand and poised to begin, he paused. He needed to start tracking down the witch.

He shaved slowly and with precision, enjoying the scrape across his jawline and cheeks. When he finished, he stared back at a handsome face and a neatly trimmed goatee. His reflection smiled back at him. He knew where to begin.

~

"I trust you will be joining us tonight, Angelica?" My mother delicately popped a small bite of game hen into her cupid's bow mouth.

"What's happening tonight?" I asked, feeling a tickle of unease creeping up my spine.

"For the Hunt, of course!" She tittered and reached to place her hand atop my father's.

"Oh... I dunno."

Absolutely not. But I couldn't voice my refusal without sparking a scene. And I was already treading carefully.

"Why the hesitation?" My father demanded, tearing his eyes away from my mother's lips. "And speak properly. Lazy speech is sickening." My father sneered and grabbed his goblet next to his empty plate. Why he bothered with a setting at all never made sense to me.

"I'm feeling a bit fatigued from the drive. I was actually considering turning in early," I replied, though it wasn't true.

I had not been, but if it meant avoiding the hunt, turning in early it would be. Kaden watched the exchange quietly. He knew better than to interject.

“You're coming, and that's final, young lady.”

“But-”

“No.”

He twirled the stem of his glass in his fingertips. “No excuses, Angelica.”

Preparations for the Hunt commenced just after dinner. My blood ran cold when the hunted were paraded out for selection. The first was a young man, emaciated and adorned with deep wounds on his shoulder and wrists. He stood before us, trembling, clad in nothing but torn pants and an iron choker with a single O-ring in the center. My mother cast a disdainful glance at him and dismissed him with a small moue of her lips. Another was brought forth, then another. Each in the same wretched state, each consumed by terror.

But when a woman was presented, my heart plummeted. I knew my mother would take interest in her, and I mourned for the poor girl instantly. Though gaunt, her beauty was undeniable. I surmised she was one of the unfortunate slaves my father indulged in beyond mere sustenance. Through no fault of her own, she had sealed her own fate. A cruel smile spread across my mother's face.

"She's perfect."

The girl fell to her knees with a cry. Two servants silently lifted her by her arms and carried her off. My mother turned on her heel, trailing her hand across my father's chest as she passed him. He licked his lips and followed her. I shuddered and faced Kaden.

"We have to help her."

"How, Pepper?" He cast a look in the direction of my retreating parents.

"Don't worry about them." My stomach turned over. "They'll be occupied for a while."

"Oh." He made a face. "Oh, gross."

"Tell me about it." I grabbed his arm and started dragging him toward the door the servants had exited with the girl.

"Wait, wait, wait, Pepper." He yanked his arm out of my grip.

"What? You want that poor girl to be torn apart by the hounds? My father? Worse, my mother?"

"Of course not. But we have to be smart about this." He stopped walking and faced me. "We can't just go storming the gates, fight off an armed livery, and rescue the maiden."

He leaned in, whispering urgently. “We get her out of here using the Hunt as cover.”

I took a deep breath. “Right. You’re right.”

“Let’s go to your room and plan.”

“Can’t. The walls have ears.” I gritted my teeth. “My mother has the entire house warded, so she can keep an eye and ear on everything that happens. I can’t guarantee the garden is clean, but it’s our best bet for privacy.”

“Then I suggest we take a walk.”

~

Damien dabbed at the corners of his mouth and folded his napkin in front of him. He motioned for the waiter to pour him some more wine, who complied in halting movements.

“Don’t fight it. It’s easier if you just submit.” Damien nodded to the bodies of the waitstaff behind him. “Learn from your compatriots’ mistakes.”

The waiter’s eyes shone with tears, and Damien rolled his.

“Don’t be sentimental, you fool. All of your lives are match flames in a windstorm. In fact…” He turned his wrist, and the waiter's head spun on his neck. "I have no time to waste educating fools."

He lifted the wine from the still-standing waiter's hands and topped off his glass. He sniffed it appreciatively. "A good recommendation, however. It's a lovely Charbono."

~

The hounds strained at their leashes. Bloody foam dripped from their jowls, and their eyes sparked with flame. Hairless and muscled, a few winked in and out of visibility. My father's hellhounds were terrifying, even to me.

The young girl stood bound in front of them. She still wore her tattered dress, and her quaking body looked even smaller than it had before. Though unnecessary for the hounds, Mother always made a spectacle of allowing them to get the scent. I carefully schooled my expression to keep the disgust and fear hidden. I couldn't risk any attention falling on me.

Mother strode up to the girl and lifted a ceremonial athame. She slid the blade between the girl's hands and cut the ropes.

"The Hunt has begun!" she cried to the sky, and the girl started running. They always run. If I could, I would have told her that running only whips the hounds into more of a frenzy. My mother, too. I heard my father's deep chuckle behind me. He was thinking the same thing as I. I didn't get the joke.

I watched her stumble as she tried to run, falling twice before she made it to the treeline. Every time she fell, my mother would giggle. She scratched behind the ear of the lead hound. They always gave the Hunted twenty minutes to get a head start. I could hear the blood rushing in my ears, and my face felt cold.

There was more riding on our plan than just saving the young woman's life. If my parents were to discover my duplicity, the best-case scenario would be their refusal to help with Damien. The worst-case scenario was fatal. I glanced over at Kaden and saw my fear mirrored in his eyes. The minutes crawled by. Finally, my mother lifted her arms and dropped them as if beginning a drag race. The leashes on the hounds disappeared, and they charged toward the treeline, letting loose their bloodcurdling howls. Their baying always reminded me of what a thousand tormented souls would sound like. It made my skin crawl.

We all began to follow. Kaden and I gravitated toward each other, sweating in the wake of the hounds' fire-scorched path. I could feel the heat of the earth through the soles of my shoes as we trotted after my parents.

There was no way this was going to work. I almost turned to Kaden to tell him we needed to call it off when I heard the hounds baying in the distance, and the words died in my throat. They had the scent and were homing in. Shit. We were out of time already. I saw my mother grip my father's hand excitedly as she hopped up and down like a little girl.

Just like that, our plan was in jeopardy. We had counted on the girl being able to get a little distance or hide before the hounds caught up with her, but it seemed she was feeling either too exhausted or too defeated.

My mother dropped my father's hand, and he stood off to the side, watching her with a smile. She chanted in low tones, and a wind began to kick up. She grinned, her eyes rolling in her head, while her skirts billowed and her hair whipped around her like rabid snakes. A lightning bolt struck the trees, blasting them thirty yards back. Another struck, and more trees flew. We could just make out the movement of the hounds in the darkness, and Mother bolted, my father loping to catch up.

I felt like crying with the helplessness of it all. I needed these two in my corner to contain Damien, but I didn't want them anywhere near me. Add to that my inability to help some poor girl from a cruel and senseless death, and my feet stopped moving.

Then I got angry. The real power my parents had over me had nothing to do with hellhounds, sharp canines, or magic. They made me forget myself. I became less than nothing around them and reverted to the little girl who spoke softly and tried her damndest to remain invisible.

I started running and smacked Kaden on the back as I passed him.

"C'mon! It may not be too late after all!"

"What?? Are you crazy??"

"Yes! C'mon!"

I ran as hard as I could until my breathing burned in my chest. I passed my father and almost laughed at his surprised expression. I reached my mother and matched her stride.

"Angelica? What are you doing?" Her face furrowed with confusion before it lightened into tentative hope. "Have you finally chosen to join us? Rache!" She called over her shoulder, "Our little girl is taking point on the Hunt!" She clapped her hands in delight, and the joy in her face was enough to pull at my heart a little. Every child wants the approval of their parents. Even if they are monsters. I passed her, and she cheered.

I vaulted over felled trees and zigzagged boulders and branches. Leaf litter shuffled beneath my flying feet. If I could just get a little more distance, we may have a chance. I barked a short scream when Kaden appeared next to me.

"Jesus, you ass! I almost zapped you!"

"By the gods, Pepper. How long have we known each other? And you still let me startle you?"

"Shut up and move faster. We need to get a better lead before you double back to distract them."

"Yeah. Tell me why I have to be the distraction again?"

"Because they like you and won't care if I'm hurt."

"Damn solid logic. Okay. Let's stop talking because I can't run as fast as it is, and talking is making it worse."

I made the motion of zipping my lips and focused on the path ahead.

The hellhounds had become ominously quiet, which usually meant cornered prey. I put on a final burst of speed and shot into a clearing. The girl had her back to a tree and was surrounded by the snarling hounds. Their muscles rippled in the shadows, and the ground sizzled where their spittle fell. Her knees buckled when she saw me, and her eyes widened. Desperate tears ran down her face.

“Please don’t kill me. I’ve done everything asked and then some. I’ve done nothing wrong.”

I sighed. “You have no idea the dynamic of this miserable family if you think pleading will get you anywhere.”

I took a step toward her, and she backed up further against the tree.

“Look, I’m not here to hurt you. What is your name?”

“Estelle.”

“Okay, Estelle. We’re going to try to get you out of here, but you have to trust me.”

“What?”

“Jesus, girl. I’m trying to help you. You have to trust me!”

"O-Okay."

I raised a hand and called the hounds to me. They hesitated for a moment, confused, before trotting up to me. They were nothing if not obedient. I swallowed my fear and crouched down to pet the lead hound. I looked up, but Estelle still hadn't moved. I bit back my frustration and said,

"For God's sake, go, you twit."

She started running, and the hounds snarled to attention. I threw up a hand, halting her in her steps.

"Don't. Run."

She nodded and walked quickly toward the edge of the clearing. I blew out a breath and closed my eyes. I concentrated on her scent, trying to erase it. I linked with the hounds' minds and worked on breaking the connection to her.

Just as I felt them begin to respond, my parents strode into the clearing to find me crouched down, my hands caressing the face of the lead hound.

"What's happening here, Angelica?" My father narrowed his eyes on the scene in front of him. I swallowed hard. This was it.

"They lost the scent. I was just consoling the big guy here."

"They lost the scent?"

"Apparently."

He strode up to me and grabbed me hard by my arms. He lifted me until my face was hovering inches from his, my feet dangling above the ground.

"Hellhounds don't lose a scent, Angelica. And you can't stand them." His pupils grew until they bled into the whites of his eyes, and I was looking at the face of a true demon. He threw me against a tree, and my shirt tore when I collided with the rough bark.

"You are a disappointment. And no daughter of mine." He turned to my mother. "Do it."

"Yes, love."

She strode up to me and backhanded me so hard her ring split my cheek, and my head bounced against the tree. The wind began to swirl around her again, and she began tracing runes in front of her. The roots of the tree snaked up and around me, fastening me to the ground. She lifted her arms high and began chanting. I squeezed my eyes shut, knowing these to be my last moments. I felt everything around me. The smoldering ground beneath me, the cut on my cheek. My heart was pounding in my ears, the shallow breaths in my chest. I could smell the leaves and the nearby stream. I could practically hear the music of the spheres and opened my eyes to the stars overbright above me. I saw my mother in all her fury, arms raised and wind whipping around her.

Instead of hitting me with the bolt of lightning she conjured, trees to my left exploded again. Then again. My mother lifted one hand and squeezed the air as if to choke the life from it. A body floated above the ruined treeline, clutching at her throat. Estelle. Damn.

"Not only have you failed in your ill-conceived plan, you have failed us, Angelica. And now your little attempt at heroics has cost you the help of the two people able to stand in your corner against Damien." She pursed her lips. "If I had a mind to, I'd even be inclined to help *him* as I empathize with him far more than I ever have you. I just don't care enough, nor do I think he needs any help. He's going to destroy you this time, Angelica."

Estelle floated gracefully to hover in front of me, close enough I could see myself mirrored in her wide eyes. My mother sauntered up to peer into her face.

“Hm. I suppose I see what intrigued your father. She resembles me somewhat.” She smirked, then looked at me.

“But not for long.”

Estelle’s screams could be heard from the house.

Mirror, Mirror

Draco witnessed the entire ordeal. The phoenix blood he had obtained from Morgaina earlier that day enabled him to cast his voyeur spell. There was only ever one person he watched faithfully. His heart seized in his chest when Wren struck her and practically exploded out of his chest when the lightning struck. He watched, scowling and breathless, as Wren tore apart the girl and turned to Angelica. He only began to breathe again when he saw her friend, Kaden, materialize and grab her before disappearing.

It was an ironic mercy her parents were so self-involved. Otherwise, they would be hunting her down to tear her apart, just like the girl. He believed Wren when she said she didn't care enough to help or hinder. Having spent decades with Angelica and witnessing her parents' twisted relationship firsthand, he was well-acquainted with the horrifying dynamic.

Trouble?

"I believe so, Mischief." He sat back in his chair. "I think the time has come to intervene."

That's a bad idea, friend.

He nodded absently and rubbed his chest. He was not thrilled about the prospect. In fact, he was almost certain it was suicide. But, he realized with shock that he was willing. Eternal life without her was only eternal.

"Problem is, she may just be proud enough to reject my help."

His gut twisted as he was reminded, yet again, of his unfaithfulness leading to their breakup. Decades and his heart still ached with regret. He could still clearly see the look of betrayal and heartbreak on her face and hear the shock and hurt in her voice.

She'll die, Mischief said in her simple cat logic.

"Then so will I, Mischief." He stroked her head pensively.

~

Kaden deposited me on my childhood bed with a bounce.

"Get your things. We're out of here. *Now*."

"Don't have to tell me twice." I grabbed my suitcase and my toothbrush, and he and I raced to the garage. His car alarm chirped, and we flung the doors open, collapsing into the seats with a rush of wind. We slammed the doors, and he gunned the engine, peeling out with a squeal of tires. Soon we were back on the road and able to breathe enough to talk.

"They won't come after me."

Kaden reached over and placed a comforting hand on my knee.

"I know, Pepper." He gave my knee a little squeeze. "I'm so sorry."

The fear hit me suddenly, and I burst into tears. Deep sobs wracked me as thoughts of my parents and our narrow escape crashed down on me.

"Aw, Pepper. Damn. It's gonna be okay."

"How, Kaden?" I said with hitching breaths, "My mother just tried to kill me, and now we have no resources to help with Damien, who is out there somewhere, plotting to finish the job. Ugh."

I swiped at my eyes with the heels of my hands.

"This just sucks every way to Sunday."

"You're right, it does."

He watched the road in silence for a while.

"Just try to close your eyes for a bit, Pep. Everything will still suck in the morning, so you may as well be well-rested for it."

I gave a sniffling laugh and reclined the seat as far as it would go, watching the scenery fly by.

"Thank you, Kaden."

"I'm here for you, Pepper."

"I know." I wiped the last traces of tears from my eyes and sniffled one more time.

"Thank you."

The next thing I knew, Kaden was gently shaking me awake in front of my apartment.

~

The supple leather of the Dodge Demon felt divine on Damien's skin. The growl of the 840 hp 770 ft/lbs torque V8 vibrated his chest, sending shivers up his arms. He sat with his eyes closed for a moment just to enjoy the sensations. *Better than a mint for after dinner*, he thought with a chuckle. The original owner sat in the passenger seat with a vacant expression. Damien still felt the thrill of consuming that one. *The things he did to get this car. And the house. And the bimbo with the big tits. Almost depraved enough to be admirable.* He slapped the man's belly with enthusiasm.

"And I'm so glad you told me what this car is called. I probably wouldn't have taken it otherwise, but the coincidence was too good to pass up!!"

Damien was feeling better than he had in over sixty years.

~

I sat in my window seat, staring out at the city. The failure at my parents' house was a devastating blow, and I was wracking my brain to try to come up with something, anything, to compensate for the strength my parents' help would have given me. I consulted my cards over and over.

And I kept coming up with nothing.

Well, almost nothing. I had come up with one option, but it was a tremendously unsavory one. He was one bastard

I was not eager to reach out to. I rested my head against the window and sighed.

But there it was, the last option. The only one which was only slightly less repugnant than reaching out to my parents. And if I could face my parents and survive, surely facing the man who broke my heart was something I could handle as well. It just stung. A lot.

I got up from my seat at the window, retrieved my phone from the arm of the couch, and dialed Kaden.

Caught Between A Bad Plan And No Plan At All

"Oh, Pepper."

"I know."

"We have to think of something else."

"I know."

There was a pause on Kaden's end. "Is there anything else, anything at all, we can think of?"

"Nope," I said with a sigh. "Believe me, K, the idea of reaching out to him is giving me an ulcer. But I can't think of anyone else. Despite our history, he would be the strong backup we need in this fight and we both know it."

Now it was Kaden's turn to sigh.

"I wish there were another option."

"Me, too. But I can't think of one."

"Okay. Do you want me to find him or do you want to?"

"I got it." I screwed my eyes shut and my stomach flipped. Kaden was not going to like this. "I never lost him."

"Excuse me?"

"I've always known where he is, K."

"Oh."

I nodded even though he couldn't see me. Suddenly, I wanted nothing more than to get off the phone. I felt sick and depressed.

"I gotta go, K. I can't talk about him anymore."

"Righto. Just know when I see him, I'm going to kick his ass."

"Get in line."

I hung up the phone and immediately ran for the bathroom to throw up my breakfast. Damn him for having this effect on me still. I scrubbed the tears from my face and brushed my teeth before beginning a bath. Then I sank to the floor and cried.

~

Draco hung his head. Every tear she cried was his.

"Jesus, Mischief. I'm still hurting her, and we haven't talked in almost thirty years. I don't know if this is such a good idea."

Of course not. It's a terrible idea, and I'm going to end up without my friend.

He had nothing to say to that. He just shook his head and made his way downstairs to take his frustration out on a heavy bag.

After two hours of attacking his training equipment, rage and shame still sat heavy in his chest until he was throwing runes and spells at the reinforced concrete walls. Finally, exhausted and depleted, he crawled up the stairs to his bed and slept for the rest of the day.

Road Tripping

After disposing of the man's body in a nearby river, Damien sped through the countryside on his way to the Angelicus estate. He was several states away, but the car would get him there in a couple of days. He smiled and sang along with the 50s music he found on something called 'XM Radio.' Imagining everything he was going to do to the witch had him practically buoyant.

He pulled into a gas station to once again fill the tank and sat in the car, waiting for the attendant. After five minutes or so, he remembered full service was now a thing of the long past. Irritated, he stepped out of the car and walked up to the building carrying the man's wallet. Inside were plastic cards he found worked like magic to pay for things. He strode up to the attendant behind the counter and smiled at her.

"I need to fill up on pump six."

"Anything else for you, sir?"

Damien smiled wider. Her soul smelled sweet and crisp. Divine. He drove his fist through the partition and grabbed her by the neck. Plexiglass shattered and fell around her as he pulled her through. The scent of her blood made his cock jump.

"You know? I think there is just one more thing."

~

Kaden and I were once again in his car, this time headed for Los Angeles. Though it was not the first place I wanted to go, at least it was heading in the opposite direction of my parents' place. Despite being pretty sure my mother had no intention of finishing me off, it still felt good to get as far away from them as possible. And few places were further from upstate New York than Los Angeles.

"Tell me again why he would live in that place. It's so overcrowded - and with such botox-filled, superficial cretins-"

"Enough, K. It's not that bad."

"Whatever you say. How did you live there for so long?"

"It's actually a pretty cool town if you know where to go."

"Doesn't matter where you go. You're still up to your ass in people."

"I guess you just become accustomed to it. I didn't even notice after a while."

"City of Angels. Pfft."

"You're a peach this morning."

He was silent for a while, and I could practically hear the gears turning.

"I guess it's not all of the people there," he said quietly, glancing at me. "It's really just one I'm dreading seeing, and he is precisely the one we're headed toward."

I leaned my head back and closed my eyes.

"I know, K. But our options are dry."

A muscle jumped in his jaw, but he said nothing else.

~

Mischief woke Draco mid-morning the next day.

Are you feeling better?

Draco stretched and glanced at his watch. "Holy crap, Mischief. I slept fourteen hours?"

I'm not great with time, but I know dinner and breakfast passed before I woke you up.

"I'm sorry, friend. Are you hungry?"

No, made do. There are plenty of mice in the alley. None even had fleas.

"I promise to make it up to you." He swung his legs off the bed and stood.

Salmon?

"If that's what you want. Or would you prefer In-N-Out Burger?"

Tough choice. Surprise me.

"You got it."

He'd been so wiped the day before that he couldn't even remember how he'd stripped to his boxer briefs. He strolled over to his mirror to check on Angelica and stuttered to a stop, mouth agape.

She was heading his way. She was in Kaden's car and coming to him.

She was coming home.

~

The volume on the radio was a low droning in the otherwise silent car. Stress rolled off of Kaden in waves, and it was rubbing off on me, causing my own anxiety to spike.

"K, please. I understand you're unhappy about this, but your energy is making my skin crawl. Please try to relax."

"I'm sorry, Pepper," he said with a long exhale, "I'm worried for you, is all."

"I know. I'm worried for me, too. But you're not helping."

We pulled into the gas station, and Kaden put the car in park beside one of the pumps.

"Speaking of making my skin crawl, do you feel that?

Kaden nodded with narrowed eyes.

"Something's happened here. Wait in the car, and I'll go check it out."

"No way," I said, already opening my door and striding toward the building.

The scene was chilling. Snacks were scattered on the floor next to an overturned shelf. The partition separating the cashier and the front counter was a ruin of plexiglass. I could practically taste the adrenaline and fear impregnating the space. The distinctive spatter of blood marred the counter and glass. I looked at Kaden as he surveyed the scene with wide eyes.

"He was here, wasn't he?" I asked him, my voice tremulous.

"It would appear so."

"Now the question becomes, are we following him, or did we just miss him going the other direction?"

Neither scenario was ideal. Either we were chasing him, or he was heading East. I couldn't decide which situation was more horrifying. If he were heading West, we chanced to face him without backup. If he were heading East…

There was only one possible destination to the East. The Angelicus Estate. My mother's taunts echoed in my head.

If I had a mind to, I'd even be inclined to help him as I empathize with him far more than I ever have you. If that were to happen, there was nothing in heaven or on earth to save me this time, powerful mage in my corner or not.

He's going to destroy you this time, Angelica.

Tears of frustration and fear sprang to my eyes. I could tell by the look on Kaden's face he had arrived at the same conclusion. I wracked my brain to try and recall seeing any other cars on the road which might have been his, but I had been too focused on everything else. And I wouldn't even know what kind of car to look for.

"I'll call in an anonymous tip." Kaden headed back outside, and I followed him. I got back to the car, the smell of blood still fresh in my mind. I felt sorry for whoever had been unfortunate enough to have snagged Damien's interest. I could certainly empathize.

Kaden made the call to the police while I pumped the gas. It felt eerie, pumping gas as if everything were normal. Meanwhile, there was a crime scene smeared with blood no more than twenty yards away.

"Okay... Thank you. My name? It's-" Kaden hung up the call and pocketed his phone. "Should probably get a move on. The cops will be here any minute."

"Just about done."

Kaden got in the car and rested his hands on the steering wheel. He started cataloging the images he had picked up from the store: a young girl, late teens maybe, with strawberry-blonde hair and sparkling blue eyes, smiling at the handsome stranger; the picture of customer service. Her terrified expression as his fist shot through the glass in front of her. The panicked thoughts of having no time to call the police and of leaving the store unattended. The boyfriend she'd been texting before Damien walked in. Kaden shook his head.

I finished with the gas and capped the tank quickly, eager to split before the cops arrived. The thought of the time wasted answering questions had me practically leaping into the car, slamming the door after me. Kaden jumped at the sound. He turned to face me.

"I got some images. We'll discuss them along the way." He fired up the engine, dropped it into drive, and we were on our way again.

~

Draco cleaned like a man possessed by an OCD demon. Put off by his manic behavior. Mischief was through the cat door and into the alley by the time he had started cleaning the second wall.

His heart was in his throat, and he couldn't get his pulse to stop racing. Though they were still a few days away, he hadn't seen Angelica in almost thirty years. He needed to keep busy, or he would lose his mind.

He was just finishing up cleaning one of the bathrooms a second time when the phone rang. He froze. The ringing was coming from his landline, a phone which never rang. The only reason he had it at all was it was the number they'd had when he and Angelica were together. He couldn't bring himself to sever a tie she might use should she ever want to talk to him again.

And now it was ringing.

His heart kicked up another notch, and he threw the sponge into the tub before jogging to the phone, stripping off his gloves as he went. He lifted the receiver and brought it to his ear with equal parts fear and hope. His throat constricted, and he swallowed.

"Drake?"

"Yes."

"It's…"

"I know who it is, Gelli," he said softly. He heard her breath hitch on the other end.

"I really shouldn't be calling."

"I'm glad you did."

~

The sound of his voice caused a sweet ache in my chest, and my eyes blurred. Kaden reached over and took my hand. When I heard Draco's pet name for me, I nearly lost my mind. I hadn't heard that name in twenty-seven years, and it still kicked up the butterflies in my stomach. A couple of tears slid silently down my cheeks.

"I…"

This was so much harder than I'd anticipated. I muted my end and sniffled. I was just about to ask a huge favor from the only man I have ever loved and the only one to break my heart. I swallowed hard and unmuted my phone.

"Damien's loose."

"I know."

How he knew was beyond me, but I wasn't interested in going down that rabbit hole.

"I need-"

"I'll do it."

I sat back, stunned. He hadn't hesitated. Hell, he hadn't even let me get the question out.

"Are you sure? This is practically suicide."

"I've considered that."

"I would've reached out to someone else, but my parents are a no-go, and you're…"

I hesitated. I hated to stroke the man's ego, but the truth was, I had never met a stronger mage. "We're on our way to Los Angeles right now," I continued. "I had planned on asking you in person."

"This is fine, really. Like I said, I'm glad you called me."

"You're still in the same place?" I asked, though I knew the answer.

"Same place, yes."

"How's Mischief?"

"A pest," he chuckled, and my heart stuttered.

I'd found Mischief when she was approximately eight weeks old, digging through the trash. Watching her wobble and struggle to walk, I had assumed she'd been injured in an accident. After some research, we'd surmised her lack of fine motor skills to be neurological. That and she told us. She was the only survivor of a virus which had claimed all her sisters and brothers along with her mother. Draco bonded with her as a familiar almost immediately, which surprised us both. But now she was bonded with a mage—an exceptionally powerful mage at that—she'd live as long as their bond did. He spoke again and drew me back to the conversation.

"I'll start doing some research so we have a starting point when you get here."

"We should be there in a few days."

"That's fine."

"Maybe sooner."

He paused.

"Just get here, Gelli. And please be careful."

"I'll try."

Years ago, this would be the point in the conversation when we would exchange "I love you". Before the moment got too awkward, I hung up.

~

Draco dropped the phone back in its cradle and let out a long breath. The phone call had actually gone far better than he could have hoped. Over the years, he had pictured their first conversation in every way conceivable. The one they just had was pretty close to a best-case scenario.

He couldn't begin to consider she'd forgive him, and he dared not think of the prospect of her returning to his life.

All he could do was clean.

Warring In A Winter Wonderland

The car was silent in the wake of my first conversation with Draco since I'd left. Snowflakes began drifting down just as we crossed the Arizona border. The flakes grew heavier and more voluminous until we found ourselves in a virtual blizzard. Kaden drove slower and slower as visibility approached zero. He had slowed the car to a crawl until the weather stopped us altogether in Flagstaff.

"Let's stop here for the night," Kaden said, nodding toward a small, rustic-looking motel set a short way back from the highway.

"Sounds good to me. I could use a long, hot shower and a good night's sleep.

"Amen to that."

Kaden went into the office to secure a room for us while I waited in the car. He came back out with keys in his hand and a grim expression on his face. He got back in the car, and I looked at him expectantly.

"Two things: One, this weather isn't supposed to get any better for the next few hours, and we may be stuck here for a day or two. Second, the receptionist in there is not human."

"Ducky."

"I'm thinking demon, but hoping vampire."

"Damn."

Kaden started the car again and drove us around to our unit. It was the last one in the back, facing a wooded area. We got out and made our way to the trunk to grab our bags. I had packed a bit heavier this time, knowing the trip would be longer. Kaden shouldered the majority of the luggage, and I got the rest. Something moved in my peripheral vision, and I tried to catch it without giving away that I'd seen it.

"Okay, I don't think we're alone," I whispered through clenched teeth.

"Shit. I was hoping we'd have some time to settle in and prepare."

"It was just a flash, so maybe we have a little time."

"Well, let's get inside." He unlocked the door, and we shuffled in with our things.

We had just dropped our luggage when something hit the window with a loud crack. We dove for cover, and I crawled over to check the window. It was still intact. Score one for our team.

"K, close the vents."

"On it."

He scurried over to the wall, shut the vents, and dropped back to the floor. We couldn't risk leaving any leaks into or out of the room. I crouched and ran to the bathroom to grab a towel. On my way back to the front door, I unzipped one of my bags and retrieved the rock salt I always carried. I tucked the towel into the gap between the door and the floor, then poured a liberal amount of salt in front of it. Kaden was already drawing a ward on the carpet. Once finished, I tossed him the rock salt, and he lined the windowsills.

I stood and began casting runes in front of the door, warding as quickly as I dared. There was no room for error with warding spells, and I couldn't afford to make a mistake out of panic.

Just as I finished casting the final rune, something hit the door with enough force to rattle the windows. I stumbled back a couple of steps and nearly collided with Kaden behind me. We hugged the wall and made our way to the window to assess what we were up against.

Eight frost demons stood in formation, their pale blue bodies clad in icy armor and glistening armaments. Two imps struggled to drag a trebuchet loaded with giant balls of ice.

Our motel room had suddenly become a bunker. I huffed out an annoyed breath, and Kaden turned to me, eyebrows raised.

"What? I was really looking forward to a hot shower and some sleep. This bullshit is putting a big, fat dent in my happy right now."

He smiled, and it dawned on me that it had been days since I had seen a genuine smile on his face. I'd been so absorbed in my own misery, I hadn't noticed my best friend was suffering. I'd definitely make it up to him if we survived the night. Big if. They already had us on the defense, and the fight hadn't even begun.

Kaden began casting, and a warm breeze enveloped us. The cheap polyester curtains billowed, and my hair lifted. He cracked the window and pushed his hands outward. Flames shot out. The demons and one imp jumped back, but he caught the other imp, who flailed about before dropping to the snow with a hiss. The ice ballistas melted.

The demons all produced shields and dropped into a phalanx. Kaden kept the fire going, but the shields seemed to be made of something other than ice, and the demons held position. One raised his hand, and a shimmery, black fog sprang from his fingertips. Kaden hastily shut the window.

I kept an eye on the vents he had shut. Unfortunately, the insulation in the cheap motel room was poor. A thin, pearly black mist began seeping into the room.

"Kaden! Above you!"

"Shit!"

I began conjuring a containment spell. I focused as best I could on the casting and soon had the spell locked in. I closed my eyes and allowed the magic to take over. A purplish-silver mist left my hands and headed toward the fog, now halfway into the room.

Sweat beaded on my upper lip as I concentrated on pushing back the invasive fog. The demon that had sent it wasn't the strongest, and I managed to force it to retreat. I followed with my spell as the seeping fog withdrew through the vent and then worked on sealing the leak. Another loud bang rocked the room, but I ignored it, focusing on completing the seal.

When the vent was finally secure, I let go of the spell and blew out a breath. Kaden was conjuring more fire, and I followed suit.

"We may have to go out there. Shooting through this little window can only do so much."

"Let's try a few more things first, yes?" I replied, glancing at the glass.

"Definitely. Can you cast a little further?"

"What's the plan?"

"I'll blast them if you can get a rictus spell going."

"Gotcha. Good plan." I chewed my lip. "K, if they're demons, do you think they have Damien on speed dial?"

"Not sure, but if they do, I believe our travel schedule will have to be bumped up a notch, snow or no snow."

I nodded, unable to trust my voice. One terrible thing at a time.

"One thing we have on our side is that Damien is not well-liked in the hoofed community." Kaden shot another stream of fire at them. The remaining imp went down. Two down, eight to go. Goody.

"He's almost as well-liked as we are."

The thought lifted my spirits slightly. Perhaps we were just up against some demons trying to kill us for the usual reasons: our being exterminators and bounty hunters by trade.

I began casting runes for a spell to envelop and paralyze the demons. Probing their energies, I found they were all lower-level, which served to lift my spirits even further.

"They're all lackeys, K. We can take them."

My confidence restored, I drew runes with a flourish. The energy required for this type of spell typically involved working up some serious rage. I, fortunately, had it in spades at the moment.

I completed the first portion of the spell and went to work casting the second. The circle had them surrounded. Now I just needed to lock it down and cast the rictus.

Another boom shook the room as they fired on us again. This time, big enough one of the walls splintered, and a couple of ceiling tiles fell, dusting us with dirt and whatever ceiling tiles are made of. A couple more hits and they would be in. I heard Kaden gasp behind me. I glanced over and saw him removing a massive splinter of wood from his arm.

"You okay over there?"

"I'm fine. Let's just worry about getting rid of these bastards. Then I'll have time to bleed."

"Copy that."

The circle was complete. I threw a series of runes to cast the rictus spell. The demons tried to retreat but were soon stopped by my circle. I smirked and continued to cast. One of the demons bellowed with frustration, throwing his lance to the ground, leaving him with a large sword belted to his waist.

"I've got them. Hit them with as much as you got."

Kaden shifted slightly and began casting bigger. I closed my eyes, chanting in low tones to keep the spell intact, and took a step back. I could feel the moisture burning up in the room. I peeked through my lashes and watched as his eyes narrowed, the brown irises disappearing to be replaced by the swirling colors of fire. The demons stood stock-still, attempts at retreat frozen by my spell. Kaden let the fire go.

Inhuman howling split the air, and each of them dropped, one by one, onto the snowy ground. When the final demon stopped moving, I released the spell and took a deep breath.

"Nice work, K."

"Back at ya' Pep."

"Let's get you patched up, and then I call first shower."

"Deal."

Maybe I'd even get that good night's sleep I'd wanted.

All-You-Can-Eat Buffet

Damien sighed in contentment. He couldn't recall the last time he had such a consistent supply of delectable souls, and his trip across the U.S. had provided him with a veritable smorgasbord. He chuckled to himself. His newest meal sat beside him, eyes vacant. The macabre company kept his mood light, plus, for some reason, he found it amusing to drive in a carpool lane. Halfway through Pennsylvania, the car speakers trilled, indicating a phone call was coming through.

Initially, he ignored it. The man to whom the car had belonged seemed to receive quite a few calls. Once Damien figured out how the telephone worked in the car, he was able to reject the calls before they had a chance to ring twice. The speakers started ringing again and again, but he rejected them. It rang again, the same number. The third time it rang, Damien picked up.

“Speak.”

“My Lord?”

"Ah, Callisto. You have news for me?"

"I do. I tried calling yesterday but couldn’t get through."

"Yes, I have picked up a few calls. They were not for me."

"They came to my motel, sir."

Damien pulled over to the side of the road in a spray of gravel.

“Repeat that.”

“They came to my motel, sir.”

Damien slammed his fist against the steering wheel, causing it to creak from the impact. He wasn't yet strong enough to travel through the planes, and now he faced a crucial decision. It appeared that the witch wasn't with her parents. Was she on her way to their house or driving away from it? If she was heading toward her parents, he could beat her there. Advantage. He also knew there was no love lost between Wren Vander and her daughter, which meant proposing an alliance with them might work in his favor. Advantage. If she was going West, he knew her destination, but he would be chasing her. Minor disadvantage. Continuing East was the clear choice.

Moreover, it had been an eternity since he last saw Rache and his wife, Wren. Though he and Rache shared centuries of history before he adopted that ridiculous German name, he openly admired Wren. She was a constant source of entertainment and speculation among his brethren whenever he happened to be downstairs. Indeed, watching her was a favored pastime among all demons. He couldn't help but smile at the memory of one particular instance when Wren flayed a woman alive for flirting with Rache. That incident solidified her as a favorite among the rabble and elevated her to hero status among the newly initiated and lower levels.

There was speculation among the Court regarding whether she had demon blood in her ancestry or if her soul was naturally black. Damien's opinion leaned toward the latter. She demonstrated more imagination in her methods than his kin, who tended more toward rape, raze, and ruination.

Pulling up to yet another gas station, he relished the car despite its gas-guzzling habits. This time, he skipped the attendant and simply used one of the plastic cards to fill up. As the car consumed fuel voraciously, a truck pulled up on the other side of the pumps. A large man in overalls spilled out of the driver's seat and locked eyes with Damien. Damien acknowledged him with a curt nod.

The man spewed a stream of tobacco juice on the ground in front of Damien, narrowly missing his shoe. Damien dropped his eyes to the stain on the ground and then dragged his gaze back up to the mustachioed figure, trying to maintain a neutral expression. He didn't have time for this.

"Chew lookin' at, purdy boy?"

Damien narrowed his eyes and said nothing.

"I said, 'chew lookin' at, purdy boy?"

Damien shook his head and turned his attention back to the pump.

"Ya' got some purdy eyes, purdy boy."

Damien stilled.

"Purdy...*mouth*, too." The man oozed around the station and sidled up to Damien, placing a damp hand on his shoulder. Only then did Damien smile.

Maybe he had a little time after all.

~

The storm cleared sooner than predicted, hinting that it had been unnaturally induced in the first place. Kaden and I resumed our journey, engaged in a deep and lofty conversation about the virtues of cheese puffs versus curls. We had reached that stage in a road trip where reality seemed to blur, and the landscape melded into a single image. Despite our endless topics of discussion, we had grown bored, deliberately avoiding the elephant in the car. He ran his hand over his face and finally broached the subject.

"Talk to me, Pepper. Why have you held onto his number, and why have you kept track of where he's lived?"

"Neither has changed since I left."

"But you didn't have to find out. You already knew they were unchanged."

"Yes," I replied quietly.

“Why, Pepper?” I cringed at the hurt in his voice. “And why didn’t you tell me?”

“I can’t explain it, K. I’m sorry I didn’t tell you. I was ashamed.” His shoulders slumped a little at my admission. “And I’ve hated myself for it all these years. The bastard broke my heart, but he’s the only man I’ve ever loved. I don’t know…” I banged my head lightly against the window.

“Look, I’m not mad at you, Pepper,” he said in a softer tone, “but I’ll never forgive him for what he did to you. It tore my heart out to see you so devastated for so long.”

He wasn’t exaggerating. I was a shell of a person after Draco and I broke up. I hardly ate and barely slept. Depression hung around me like a noxious cloud, driving all of my friends away except the one sitting next to me. How Kaden managed not to ditch me was a real testament to what a great guy he was.

“Kaden, what would you like to do tonight? We’ve spent so much time fixated on my problems that I’ve barely had a chance to ask you how your day has been. Never mind anything deeper than that. Let’s do something tonight. You know, go out or something.”

"You do realize those are the famous last words of the ill-fated alpha teenager in practically every horror flick."

"I'm serious."

"So am I." He gave me a sidelong glance. "Okay, I'll bite. How about a six-pack and a deck of cards?"

"Poker."

"Cribbage."

"Deal."

~

Of all the religious sects, Mormons were Damien's favorite. A religion which emerged like a magician's bunny from a hat to become one of the most misogynistic and debauched belief systems on the planet. One which embraced polygamy as a basic tenet - but only for the men. No soda - until the Mormon Church bought shares of a soft drink giant. Take all that twisted stuff and sprinkle in some racism and a healthy dose of self-righteousness, and Damien was in hog heaven. They tasted like intensely sour candy. Teeth-curling and satisfying all in one bite. Even the word 'tabernacle' was fun.

Once again, Damien found himself in the company of two young women. These had been less willing than the last two. Virgins were always more suspicious of men's intentions. He welcomed the challenge.

He couldn't decide whether to chain them or tie them, so he did both. One was shackled to the wall, the other tied to the bed. Their whimpers and cries formed a sweet duet.

They were in one of his houses in the Midwest. He had several scattered across the world and he was an immortal investor. He had bought and sold them according to the market, but he'd always kept this one. Many years ago, he'd commissioned a basement with manacled chains attached to the wall, to which he was keeping the two girls. In order to protect his secret chamber, he took a note from H.H. Holmes and killed the contractor before he could potentially share the project details with anyone. Two people can keep a secret if one of them is dead.

When you lived forever, you learned to cover your assets.

The Anticipation Is Worse

Mayhaps a good training session would help.

“You may be right, little one. I can’t settle down, and that seemed to work yesterday.”

Or maybe we go get In ‘N’ Out.

“So you’ve decided?”

There never really was any contest.

“Well, you ruined my surprise, but I agree.”

Wildstyle?

“Since when do you like onions? Cats aren’t supposed to like onions.”

I’ve liked onions since I learned about ‘wild style.’

“Fair enough. Let’s go.”

One of the benefits of having a sentient cat was that car trips involved more conversation and far less peeling an animal from an upholstered ceiling. Mischief sat quietly on the back dashboard, sunning herself through the window.

“Now, would you like to go to the one in Hollywood or on Lincoln?"

Hollywood. I like the scenery. It’s like being in a tram at one of those wildlife parks, Mischief replied.

Draco laughed.

“Is that so? I’d think you’d be used to it by now.”

I am. Doesn’t make it less entertaining. Something is different every time, Mischief explained.

Draco appreciated the advent of Bluetooth. It made him seem a lot less crazy when he appeared to onlookers as if he were talking to nothing. Or himself. Of course, in L.A., no one paid attention, too busy worrying about what they looked like to others.

Draco?

“Yes?”

What are you going to do about Angelica?

"How do you mean?"

She's engaging in a fight she can't win.

The words sounded harsh, but that was just cats. And the note of worry in Mischief's voice did not escape his notice.

"Well, she'll have a better chance with me in her corner."

She mulled that over for a while.

I suppose you're right.

Draco gave her a small smile in the rearview.

I hope you're right, she said so quietly he had to strain to hear her.

"Me, too, Mischief." He trained his eyes back on the road.

Later, at the house, Draco stood in front of his mirror while Mischief daintily munched on a cheeseburger. She had also gotten a milkshake. She declined the fries.

Is In 'N' Out owned by Fae?

"I don't believe so. Why?"

Because magic can be the only explanation for how their food is so good.

"What, fresh ingredients and proprietary seasoning isn't enough reason?"

McDonald's claims the same thing.

"Touché."

The images of Angelica and Kaden were faint, as if from a faded photograph. He sighed. It was time to anoint the mirror again. It was worth the cost and effort, but the spell was high maintenance.

A knock on his front door startled him. Few knew where he lived, and even fewer would knock on the door. There was a thump and the sound of retreating footsteps. Draco approached the door silently. He flicked an iron blade from his sleeve into his hand and opened the door.

No one stood outside, but a package wrapped in brown paper sat on the floor in front of him. He crouched down to inspect the small box. No shipping or return address. He couldn't detect any wards surrounding the item, though he could feel the gentle hum of magic pulsing from it. Sensing no immediate threat, Draco carried the package to his living room and set it on the coffee table before sitting before it on the couch.

"What should we do with this, Mischief?"

Throw it away.

"Why do you say that?"

Bad magic.

"Then shouldn't we destroy it first before throwing it away?"

Mischief padded over and sniffed the paper. She shook her head and backed up.

Bad magic. Demon magic.

"Interesting." He mused, suspicious of the innocuous-looking package. "Well, here goes nothing."

He used his blade to slice the paper and open the box. Inside was a round, hand-held mirror. Draco carefully removed the object from the packaging. Beneath the small mirror was a note which read:

NOW YOU SHALL SEE

Brow furrowed, he turned his gaze back to the mirror and the swirling mist which appeared in it. Images began to coalesce into recognizable forms, and Draco went cold.

What appeared was the inside of a car. Flashes of images interrupted the bucolic scenery through the car window. His Angelica flayed alive, screamed for her life, her beautiful face marred by claw marks and slashes. The scene changed. Now it was Angelica crying, bound and gagged, being raped by the owner of the POV. Shackled to a stone wall, emaciated and dying of thirst. Being whipped and torn apart by a cat 'o' nine tails. Decapitated. Drawn and quartered. Burned alive…

The images ran faster and faster until they were a blur of blood and screaming.

With a cry, Draco threw the mirror against the wall, and it bounced to the floor unharmed.

"He can't win, Mischief. I have no choice but to help her now."

What did you see, friend?

Draco shook his head. His parched throat clicked as he swallowed. "He can't win."

~

Damien laughed out loud as he watched, through his side of the mirror, the reactions the mage had to the images he sent him. He may not know precisely where the little witch was at the moment, but he did know her last known addresses. He had banked on nostalgia to keep the mage where he had been and knew her parents weren't going anywhere. To some, it may have seemed like a gamble to bet on the mage still being in the same place, but for all who lived lifetimes, fifty or sixty years was nothing.

He parked the car in the circular driveway and crunched through the gravel up to the front stairs. He took them two at a time and rang the doorbell without hesitation. He glanced up at the moon, now full. He'd been free for three weeks. The more time it took to get to her, the sweeter the satisfaction when he could finally exact his revenge. As he had shown the mage, he had many, many ideas about how he planned to do that.

Damien thrust his hand through the man's chest, withdrew it, and took a large bite out of his still-beating heart. The servant stood for a moment, his mouth forming an 'O' of surprise before crumpling in the foyer. Damien tossed the heart into a bush.

"Damn it, Damien, he was one of the better ones," a deep voice called from somewhere in the house.

Once upon a time, he and Rache had been close. Friends, even, if either were capable of such a thing. In fact, Rache's hellhounds had been a gift from Damien in the 14th century. Many centuries ago, they could've coined the term ‘partners in crime,’ laughing their way through the carnage they wrought upon the European continent.

“He was in my way, Agilulf.” A stony silence met his comment, and the air chilled around him.

“Don’t ever call me that again.”

Damien chuckled. “I always liked that one. Not sure why you gave it up and went with a more modern German name.”

“I know why you’re here, Damien.”

“Points to the ancient vampire.”

“The answer is no.”

“Because she is your dear miracle child?”

“She’s no daughter of mine.”

“Well, now. What did I miss?”

“She’s too human.”

“Rache.” A sultry feminine voice interrupted. “I see he’s here.”

“My Lady.” Damien bowed low in the direction of the voice. They still hadn’t come out of the shadows.

"We're not interested in your little private war on the child, Damien," she said, and the door began to close. Damien stuck out his foot to stop it.

"Now, now. Would you dismiss an old friend so quickly, Cassius?" A growl could be heard rumbling in the dark. "Oh, sorry. Old habits and all that. How about… Cassie? Is that better? I know, Ermenrich! Haven't heard *that* one in a minute, I'll wager. Now," he chuckled, "enough charades." Damien flicked his wrist, and lights blazed in the house.

"Let's get down to business."

Old Friends

My nerves were in a jumble. It felt as if moths were having a convention of moths with seizures. We were at the southern border of the Bay Area and would be at Draco's door tonight or tomorrow at the latest. Traffic permitting. Not that it ever did. Therefore, I was trapped in a car with a concerned friend and a moth convention.

"Pepper…"

"Nope."

"C'mon, Pepper. Talking might help."

"Doubt it. Besides, not much to say. Freaked out. Scared. Anticipation. Moths."

"Moths?"

"Moths."

“Why don’t you do the tour guide thing? I know you want to.”

I sighed. There was something about L.A. which brought out the tour guide in everyone. Maybe everyone got a little bit of the Industry bug and felt like telling a story. Maybe it was ego. It was a strange place which sucked you in and made you a supplicant of the L.A. Empire. Its reach spanned the country. Probably the world. Everyone from Los Angeles felt it necessary to announce where they were from at the slightest provocation. I was no exception.

“Fine. There’s a waffle place we’ll pass. I’ve been there every time on my way to or from the Bay Area. Even those few times I was sober. There’s a great miniature golf course off Ventura Boulevard, where I used to go with Draco. Waterfalls and windmills, holes with multiple layers. Castles and dragons. Shares a fence with a bird sanctuary. Pretty much what you would expect from a miniature golf course in L.A.”

“Sounds fun.”

“It really was. Once we were there, a mama duck had her ducklings in a little queue behind her in one of the pools. They were so fuzzy and cute that it almost hurt. Then, a terrifying moment was when a duckling swept over one of the waterfalls. My heart was in my throat the whole time until the little guy surfaced unharmed. Just a little scared and a lot wet."

"Aww."

Damn him. This was working. I sighed again.

"I'm going to try and close my eyes for a bit. All this anxiety is exhausting."

"I got ya'." Kaden took my hand, and I felt a sudden heaviness wash over me. My eyelids drooped, and I snuggled against the door, falling into a blissfully dreamless sleep.

~

Draco couldn't sleep. He tossed, then turned, then tossed again. His mind whirred. Angelica, his Angelica, would arrive tomorrow. Finally, he surrendered, threw the covers off, and swung his legs off the bed.

He padded over to the mirror. She slept with her hands tucked under her cheek like a child. She appeared sweet and peaceful, causing his heart to thump painfully in his chest. He ran the back of his hand along the image of her face.

He needed to disable the voyeur spell before her arrival. He couldn't even imagine her reaction if she knew he had been watching her.

~

The California dawn stirred me from sleep, and my first thought was of Draco. Kaden had pulled over and was softly snoring beside me. A gentle ache settled in my chest, but instead of causing anxiety, I felt excitement and warmth. Draco had agreed to help me without hesitation.

I was also desperate to see Mischief, my little wobblebutt. As his familiar, she naturally stayed with Draco when we split, and I hadn't seen her sweet little face in twenty-seven years.

I reached for my tarot cards and idly shuffled them.

The Devil.

The Six of Swords.

The Two of Cups.

The seizure moths stirred.

"Bugger," I muttered softly to myself.

"What's that?"

"Oh, I'm sorry, K. I didn't mean to wake you."

"That's alright," he said, stretching. "I should be up and driving anyway. What time is it?"

"5:45 a.m."

"Phew, that's early. Well, I suppose we should get a move on, then."

"Kaden."

"Yes?"

I showed him the cards I'd drawn.

"Oh."

I simply nodded.

~

Rache, Damien, and Wren all sat at the dining room table in a salesman's standoff. Neither had spoken since they'd sat, and Wren was getting bored. Finally, she broke the stalemate.

"Damien-"

"Ha. I win, Rache."

"Shut up, you buffoon. I was just going to say you're not doing anything to garner our interest by sitting there like a poorly carved statue."

"Ouch."

"You have ten seconds to speak before I throw you out of our house."

"Yes, ma'am! Ooh, I love the firm hand. Rache, you lucky dog!"

"Eight seconds."

"Fine," he said with a sigh. "I happen to know you recently tried to put the girl's lights out yourself. I'm just offering my services to help finish what you started."

"Not precisely. You want us to help you end her."

"Technicality."

Rache spoke then.

"Are you so insecure in your abilities you come to us? Aren't you able to do this on your own?"

Damien grew very still.

"I can do this on my own."

"Then why ask for our assistance?"

"That is the question of the hour, isn't it? Why do you suppose he needs our help, darling?" Wren said, slipping her hand into the crook of Rache's arm.

They both looked at Damien expectantly.

"What fun would this be without involving a romp with old friends?"

They didn't smile.

"Fine. She bested me last time with two friends and a pack of wolves. I can't afford to pull punches this time."

"He fears her!" Wren's peals of laughter grated on Damien's ears. She returned her gaze to him and leaned forward.

"Your ten seconds is up. Besides, the wolves are no more, dear. You have even less to fear than last time."

"I believe she's gone back to the mage."

"Is that so?" Wren's voice brimmed with interest.

"I wouldn't have expected that."

"I always liked him," Rache said, tucking a lock of hair behind Wren's ear.

"I, as well. And he is powerful. Almost as powerful as I."

“That’s my point!” Damien shouted, slamming his fist on the table. They turned their gazes to him. Wren’s eyes gleamed dangerously.

“You would do well to control yourself, Damien,” Rache stated in a low, icy voice.

“I believe we’re done here.” Wren stood and strode out of the dining room. Rache followed suit.

“We’re not going to fight your battle for you, Damien. But I will give you a piece of advice.”

Damien leaned back in his chair, taken aback.

“Take out the mage first. Even if she wins, that will end her.” Rache started walking away from the table. Over his shoulder, he added, “You know the way out.”

~

We were there. I looked up at the building where I’d lived for years in the Hollywood Hills. I felt rooted to my seat, staring at the cheerful displays of palms and flowers in the balcony planters—bougainvillea and birds of paradise. I studied the arched windows and doorways, trying to remember why we were there. I had never thought I’d be back. The French doors on the second floor were open, and the gauzy drapes fluttered in the light breeze. I'd always loved that spot with its stunning view of the city on one side and Griffith Park on the other.

Countless times, Draco and I explored the observatory, creating earthquakes and losing ourselves in the stars. We'd pay homage to the obelisk of great astronomers, taking turns gazing through the giant telescope. I'd peer into it, feeling a swelling in my chest, something between awe and fear, like standing on the roof of a high-rise.

We'd take the trails down on the east side, commenting on the level of visibility. There were days when the smog obscured the view, plunging the city into a hazy cloak of anonymity. Then, there were days when the air was so clear we could imagine seeing our place.

It was a couple of hours before twilight, and the air was heavy with gold and the sea. The unique smell of L.A. surrounded me—not quite tropical, but not entirely city, either. Luxurious flora graced almost every manicured garden, and the palms swayed gently. Gardeners pruned hedges and turned soil, the drone of lawnmowers and leaf blowers supplying a familiar white noise.

Suddenly, I wanted a very large, very stiff drink.

"Are you ready, Pepper?"

"Um."

"Just think, you'll get to see Mischief."

Even in my near-panic, I had to smile. I was eagerly looking forward to seeing her again. I reached for my cards. Kaden sat back in his seat, hands folded in his lap.

The Magician. Tears sprang to my eyes. Kaden looked over and locked eyes with me.

"I guess you have your answer, don't you, Pep?"

"I guess I do."

I tucked my cards away and opened the door.

Draco watched the car idling at the curb with an unusual sense of calm. When the passenger door opened, he walked to the front door and waited in the foyer. Soon, he heard footsteps approaching.

She's here

"Yes"

Mischief remained quiet for a beat.

I've missed her.

"Me, too."

The footsteps paused just outside. Draco opened the door, only to be greeted by a fist colliding solidly with his face. He staggered back, and the door swung wide.

Draco dabbed at the blood under his nose. "Nice to see you again, Kaden."

"And you, Draco."

My heart seized in my chest. He looked the same. No, better. His hair was a little longer, and he'd put on some muscle. I blew out a breath.

Mischief came padding over, her back end struggling to stay upright. She was wearing a rhinestone collar I'd given her as a joke decades ago.

Hello, Angelica. I wore my collar just for you.

I could hear the smile in her voice. She came up to me and placed her front paw on my shoe. I smiled and bent down to pick her up. She was purring softly.

I've missed you.

"I've missed you too, Mischief."

"Hi, Gelli," Draco said softly. I dragged my gaze up to meet his.

He's missed you, too, Mischief added.

Hurt and anger battled with another feeling I refused to acknowledge. I nodded, not trusting my voice. Draco closed his eyes and took a deep breath.

"Please come in," he said, taking a step back. "Unless you'd like to punch me again, Kaden."

"Maybe later."

"It's a date."

Winding Roads

Damien drove like the devil was on his tail. The engine roared, echoing the fury in Damien’s chest. Until now, he hadn't realized he had been hoping for support from the Angelicus pair. He narrowed his eyes and pressed harder on the accelerator. Blue and red lights flashed in his rearview mirror, and he chuckled.

He considered leading the policeman on a merry chase but opted for something a bit more entertaining. He slowed the car and pulled over to the shoulder. He was in the middle of Nowhere, U.S.A., and hadn’t seen another vehicle for at least twenty minutes. He didn't care enough to guess where the patrolman had come from.

Damien watched the officer open his car door and step out. He unsnapped his holster but didn't draw his gun.

The officer's ambling figure grew larger in Damien's rearview mirror until he was too close to be reflected. Damien rolled down his window.

“Hello, Officer...Chatham,” Damien greeted him, reading the name on his badge.

“License, registration, and proof of insurance, please.”

“Hm. Well, I don’t have any of those things, Officer Chatham. Registration and insurance are in the glove box, but my name isn’t on them.” Damien smiled at the surprise on the patrolman’s face. The officer's hand dropped to his gun.

“Are you saying this car isn’t yours, sir?”

“Yes, that’s precisely what I’m saying.” He pursed his lips. “Well, I suppose it belongs to me now.”

“Step out of the car, please.”

“I’d rather not, as these seats are exceedingly comfortable. But as soon as you tell me why you pulled me over, I’ll be happy to oblige.”

“I pulled you over because you were going 134 mph in a 70 mph zone,” he replied, impatience creeping into his voice.

“My, that is fast.”

“That alone could earn you some jail time.”

“Oh, dear.”

“Adding to that your lack of valid identification and admitting to driving this vehicle unauthorized gives me enough reason to ask you to step out of the car.”

“As you wish.”

Damien shut off the engine, pulled the emergency brake, and opened the door, hiding a smirk. He stood in front of Chatham and wetted his lips.

“Please keep your hands in plain sight, turn around, and place them on the vehicle.”

“I’d rather not.”

“Turn around and place your hands on the vehicle now.” Chatham’s voice had taken on a hard edge, and his hand twitched on the butt of his gun.

“Why don’t I just kill you instead?” Damien replied casually. Officer Chatham whipped his gun from its holster and fell back into a shooting stance.

“On the ground now!”

Damien started laughing. He had to admire Chatham's composure despite the fear rolling off the man in waves. Damien took a step forward, and the report of the gun split the air. A bullet hit Damien in the chest, absorbed briefly before being spat out, clattering to the asphalt.

"Why, Officer Chatham." The smile fell from Damien's face. "That was rude."

"Wha—?"

Damien growled, and his eyes wept viscous black rivulets. Chatham froze in fear, his eyes trapped by a demon's gaze. Damien's leathery wings broke through the skin on his back, stretching high and wide, blocking the sun and casting the officer's face in shadow. His obsidian eyes gleamed, and his teeth sharpened to points. Tendrils of black and red snaked around his features. He reached up and dug his taloned fingers into his scalp, peeling off his meat suit, tearing it down the middle. Thick, multifaceted, ridged horns sprouted from his forehead and curled back. Chatham peered into Damien's face and felt the beginnings of madness.

When he spoke again, Damien's voice, deep and rattling, seemed to come from everywhere and nowhere at once—a voice dragged directly from the depths of the abyss.

"Suddenly, I'm feeling a little peckish."

He struck, relishing the echo of Chatham's abbreviated scream.

~

Draco led us through the house with Mischief trotting alongside. He escorted us to our rooms to deposit our belongings, then showcased the changes he had made to the place. The most significant one involved converting the wine cellar into a training room. He clearly took full advantage of it every chance he got. His shoulders had broadened, and his waist had trimmed down. I had admired his body before, but now I was practically salivating over a physique begging to be explored in every way imaginable. While he guided us, I tried to distract myself from my lust by reacquainting myself with a home I had lived in for decades. It almost worked.

After a dinner interspersed with light conversation and awkward silence, we retired to the library to begin our research. There we sat, surrounded by books. In days past, this would have been my bliss. This had been my room. My sanctuary. Draco and I used to sit together by the fireplace in our individual literary worlds and in our shared love. My heart was beating so hard I couldn't focus.

The sight of him was intoxicating, and I had to focus hard on not looking at him. He surrounded himself with stacks of books, holding one in his hands. If I hadn't been sure I was awake, I would've sworn I was having an erotic dream. A supremely frustrating and unsatisfying erotic dream. Every time I caught him looking at me, I blushed like a damned teenager.

I spent the day oscillating between crippling anxiety, heartache, and self-loathing for wanting to jump on him.

"I need a break," I announced, rising from my seat.

Maybe a little breather would allow me to collect myself. I left the room and made my way to the closest bathroom. The closed door muted the outside noise and I felt my anxiety level drop a little.

~

"You're not off the hook, Winter." Kaden glared daggers at Draco.

"I know, Kaden." Draco slumped in his chair and sighed heavily. "Gods, she is so beautiful," he said softly, "even more than I remember."

"Too little, too late."

"Believe me, there is nothing you can say I haven't already beaten myself over the head with for twenty-seven years." He scrubbed his face with his hands. "And now she's here, and I don't know how to behave around her."

"You're not getting near her again." Kaden slammed his book shut. "Because then I'd have to kill you." He set the book aside. "Look," he said, leaning forward, "the only reason we're here is because we have no other option. Do not, for one second, believe there is any way you are getting close enough to hurt her ever again." Kaden stood and began pacing. "Jesus, Draco! You have no clue what you did to her. She didn't eat or sleep properly for years. Years!"

Draco dropped his head and nodded. He did know. He had watched the shell of his Angelica listlessly going through the motions of being alive. It had seemed she would never feel joy again. She'd refused food for so long Kaden had threatened her with a hospital stay and a feeding tube before she would eat. She'd spent days without sleep, then slept for days. She barely got out of bed. Draco witnessed it all.

"I know, Kaden."

"How could you know?? You have no idea!"

Draco paused. “I saw everything,” he confessed.

“You what?!? How?!?”

He took a deep breath. “There is a mirror in my quarters. I spelled it with a voyeur spell.”

Kaden regarded him with open-mouthed shock.

“You spied on her.”

“I had to. I knew if I never saw her again, it would end me.”

“You’re such a selfish bastard,” Kaden said through gritted teeth. Then he threw his hands in the air. “I’m going to bed.” He strode through the doorway of the library, slamming the door as he left.

Draco dropped his head into his hands. He had known this was going to be difficult, but Kaden’s ire only drove home the point he had messed up beyond fixing.

~

I splashed water on my face and washed my hands. Staring into my reflection, I noticed my dilated pupils and stubborn spots of color high on my cheeks. My hands trembled, and my fingertips tingled with adrenaline. I steadied myself on the sink. I could do this. I had to do this.

Leaving the bathroom, I returned to the library. Opening the door, I found Draco sitting alone with his head in his hands.

"What's going on? Where's Kaden?" I asked, entering the room slowly. He raised his head.

"He went to bed."

"He did? Why?" My throat went dry.

"He couldn't stand to be in the same room as I any longer."

"I see. What did you do?"

I couldn't bear to be alone with him. It was too much. Too soon.

"Just—"

"You know what? Never mind, I asked. I don't really care."

"Gelli—"

Fury surged within me.

"And you don't get to call me that anymore, Draco. You gave up that right when you betrayed me."

He swallowed whatever he was going to say. I turned on my heel to leave. The room had suddenly become far too small.

"Angelica."

I halted in my steps but didn't turn around. I heard him take a deep breath behind me.

"I have done many terrible things in my time on this planet. Lied, cheated, tortured, and killed." I heard his footsteps whisper on the carpet.

"But I have never done anything more horrible than hurting you like I did."

I turned and slapped him as hard as I could. My stomach soured, and tears sprang to my eyes. He didn't even flinch as my hand connected with his cheek. My breathing became shallow, rapid bursts of air. His eyes reflected pain, and I almost slapped him again.

"You don't get to say things like that, Draco. I won't tolerate your eleventh-hour confessions. Just because I've asked for your help doesn't mean I will ever forgive you."

"I'm not asking for your forgiveness. I know I'll never deserve it."

"Damn right."

He nodded. He stood close enough that I could feel his body heat and smell the scent of rain-soaked earth and fresh air. I shook my head.

The truth was I missed him. I missed him with every fiber of my being. My body ached with it. I wanted to close the gap and fall into his arms. Fall into him. A tear slipped out, and I shook my head again. I would never forgive him.

But I still loved him.

His brow furrowed at my tear, and he took a half-step toward me. I turned and strode out of the library. This time, I didn't stop when he called my name.

I managed to make it all the way to my room and close the door before breaking down. I slid to the floor and wept bitter tears, my heart a screaming void. I crawled to the bed and curled into a ball against the pillows, my arms wrapped around my stomach. When the pillowcase became wet, I flipped the pillow over and soaked the other side. Then I threw it against the wall and grabbed another.

I don't know how long I lay there. Just when I was beginning to drift off, there was a knock on my door. Kaden opened it and poked his head in.

"Pepper? May I come in?"

I nodded mutely.

He stepped in, closing the door behind him before crossing the room to my bedside. He sat on the edge of the mattress.

“I’m so sorry, Pepper.”

“Why are you sorry?”

“Because you’re in pain, and I hate to see you hurting.” He rubbed circles on my back. “Want me to kill him for you?”

I choked out a laugh. “Maybe after we defeat Damien. In fact,” I said, propping myself up on my elbow, “maybe Damien will take care of that for you.”

But even the thought of something that awful had my eyes blurring again. I flopped back down onto the pillows.

“How can I help, Pep?”

“Just stay with me a minute. I’ll be fine.”

“Anything you need?”

Yes. But there was no way I was going to tell him what I needed, which was for Draco to be here with his arms around me, nuzzling the back of my neck.

Kaden was quiet for a minute.

“You still love him.” A statement, not a question.

“Yes,” I replied in a small voice. “I’ve never stopped, K.”

The confession caused another wave of tears.

“If I don’t stop crying, I’ll be out of pillows.”

“I noticed the other two on the floor over there.”

“Do you hate me?”

“How could you even ask that, Pepper?” He shook his head. “That’s self-pity talking, and there’s no time for that.”

He stood and pulled the covers down, indicating I should climb in. He tucked the blanket around me and brushed a lock of hair from my forehead.

“Why are you so good to me, Kaden?”

“Because you’re my best friend and I love you, dumbass.”

I gave him a sleepy smile. “Thank you, K. I love you, too.”

My eyes were already drooping. Kaden placed a hand on my forehead, and I immediately fell asleep.

“Goodnight, Pepper.” Kaden stood and left the room.

~

Draco sat against the wall, watching the mirror. He observed her sobbing and writhing in heartache. He witnessed pillows flying across the room. When Kaden entered, he smiled slightly. He would always take care of her. After she fell asleep and Kaden left, he stared at the floor between his feet.

She still loved him. He scrubbed at the tears blurring his vision. He had no right to cry. Mischief came over and placed a paw on his knee. He stroked her downy fur and found some comfort in her soft purring.

I'm sorry, friend, he murmured.

He almost asked why until he realized he'd just heard the same conversation. He smiled at the little tabby and shifted to scratch behind her ears.

"She still loves me, Mischief."

She does.

"What do I do?"

There's nothing you can do right now except give her the space she needs. She craves you but fears you. She's confused and hurt. If I were to encounter a creature in the same state, I would give it space until it either healed or was more clear-minded.

"Sage advice as always, my friend."

Are we sleeping on the floor tonight?

"No," he chuckled and got up, gathering Mischief in his arms. He made his way to the bed and placed her on her favorite pillow before pulling back the comforter and laying down.

It was hours before he slept.

~

Damien savored the last traces of the officer's blood on his fingers. He gleefully envisioned the reactions of the other patrolmen when they discovered the abandoned cruiser drenched in blood. He imagined the bone-chilling fear they would experience when they failed to locate a body. The dispatcher's frantic voice blared through the radio.

He adjusted his suit jacket, straightened his shirt cuffs, and eased back into the driver's seat of the Demon. Cranking up the volume on the 50's music, he drove off, laughing.

Old Friends Part Deux

It's time for breakfast, Angelica.

I groaned and turned to face the wall. There was no way I was leaving this room. My eyes were so badly swollen my vision had narrowed to slits. My face felt puffy, and my head throbbed.

Are you feeling better?

The compassion in her voice prompted me to turn around and sit up to face her as she climbed her way up the bed. A cat showing sympathy was a little disorienting. She even placed a paw on my knee.

"The real question is, are you feeling okay? Everything okay?" She removed her paw from my knee and cleaned it thoroughly before answering.

He is miserable. I hate to see my friend sad.

My heart thumped traitorously in my chest.

"Yeah, well. He deserves to be miserable, don't you think?"

I do not understand.

"He slept with someone else, Mischief."

So?

It occurred to me that, being a cat, the concept of monogamy and the nature of my pain would be confusing to her. I tried a different approach.

"He betrayed me, Mischief."

That I know. But it was so long ago now. You are unhappy, and he is unhappy. You could both be happy again.

Cat logic and its oddly Confucian beauty.

"I'm not leaving this room looking like I do."

She sighed and made her way to my face in halting steps. I picked her up halfway.

Put me on your chest, please. Close your eyes.

Once she settled on my chest, she extended her front paws and placed one on each eyelid. Her feet felt silky and cool. Light pulsed bright and dim, hot and cold. Her aura washed over my skin.

Better?

I raised my hands and felt my face. Gone were the puffiness, the headache, and the painfully swollen eyelids.

"You healed me."

She nodded.

"When did you learn to do that?"

I have been learning since shortly after you left.

"Thank you, Mischief. You didn't have to do that."

Of course, I did. Otherwise, you would not have come out for breakfast, and my friend would still be sad.

"That's the only reason, then?"

She cleaned her other paw.

No.

With that, she leaped from the bed, almost sticking the landing without a wobble.

I took my time in the shower. I knew I was procrastinating, and I didn't care. Let him wait. Better yet, let him be gone before I got to the kitchen.

~

She'll be down once she finishes her bath.

"Thank you, Mischief."

"What did she say?"

"Angelica will be down soon. She's in the shower."

Kaden placed his fork on his plate. "I apologize for my behavior last night."

Draco was stunned. He had not been expecting an apology.

"Thank you, Kaden. But you really don't need to apologize. I deserve it and more."

"No, I know you've punished yourself more than that. But you understand."

"Of course I do. In fact, I have been meaning to thank you for taking care of her all these years. You're an amazing friend."

“She’s worth it.”

“Yes.” He dropped his eyes to his plate. “Yes, she is.”

~

I made my way to the kitchen and heard laughter. When I turned the corner and entered the room, I hadn't expected to see Kaden and Lucien leaning toward each other and giggling like girls at a slumber party. Honestly, I had pictured more blood flying.

“There she is,” Kaden called out. “Good morning, sleepyhead.”

“We were just talking about you,” Draco said with a wink that made my stomach flip.

“And laughing?”

“Yes, well…” Kaden blushed lightly. “We were reminiscing about funnier moments.”

“Involving me?”

“More… about you, I guess.”

“Nice. Making fun of a girl while she’s showering. Pretty sure you’re not supposed to do that.”

“Sorry,” they said in unison.

"Asshats."

The spread looked delectable. Bacon was piled high next to buttered toast, pancakes, steaming muffins, and eggs. Fresh fruit was cut into bite-sized pieces, and a carafe each of orange juice and grapefruit bookended the meal.

"Kind of a lot of food for three, don't you think?"

"Um... yes?" Draco cleared his throat. "I know breakfast is your favorite. I didn't want to run out."

"You really think I could eat all of this?" I pointed my fork at him. "What kind of bottomless pit do you think I am?"

Kaden and Draco exchanged a glance before fixing twin wide-eyed gazes on me.

"Stop looking at me with that tone of voice. I'm not-" I threw my hands up. "Just shut up." Draco chuckled and took a sip of juice.

"Any luck finding something to use against Damien?" I asked Draco, gesturing to Kaden. He handed me the platter of bacon.

"I did find some interesting spells after you two went to bed. I'll show them to you once we're finished here."

He cut a piece of toast with his fork and dipped it in his eggs. I munched on bacon. Kaden cleared his throat.

"Well, I'm going to hit the shower," he said, and I shot him a panicked look. He couldn't leave me alone with Draco. He wouldn't. He smirked at me, shrugged, then turned and left.

Asshole.

Even though the moth convention had reconvened, I grudgingly admitted to myself that Tweedledee and Tweedledum were right. Very little could dampen my appetite for breakfast foods. I made a little sandwich of bacon and eggs, using syrupy pieces of pancake. I glanced up at Draco and found him watching me with a smile. I narrowed my eyes at him. He quickly lifted his coffee cup to hide it, but his eyes still twinkled.

"I think I found something which may help," Draco said after swallowing a sip, "it's similar to a containment spell, but…"

"But?"

"We need some muscle, Angelica. Two mages and a witch are not enough. We're missing brute strength."

"Gee, Draco, I hadn't considered that." I started stabbing at some fruit on my plate.

"Why don't I just call up my friendly wolf pack? Oh, wait. That's right. They aren't a 'pack' anymore. They are an endangered species because there are only two left. And they won't speak to me because I managed to get every other wolf in their family killed the last time." He laid down his fork.

"I know that Gelli-"

"What did I say about that name?"

"I apologize. Angelica. And I have an idea about that, too."

"Pray tell."

He sighed and pushed his plate away. I snagged one of his remaining pieces of bacon.

"Golems."

I raised an eyebrow. It was actually a great idea.

"Interesting," I replied.

It had not been an option I had considered. I was not versed in the craft, but I was damned willing to learn now.

“We need an Ancient. At the very least to, teach me, if not do the work altogether. And they are not known to be especially sociable,” I pointed out.

“No, they’re not. But…” he trailed off, and I huffed in annoyance.

“You have to start finishing your sentences, Draco. Time is a factor, you know.”

This earned me a smile.

“I know someone who can get us in touch with one.”

“Really?” I felt the smallest tickle of hope before I quashed it. There was no room for hope yet.

“Yes. I can see her today, in fact.”

I sat back in my chair and chewed my lip.

“Draco.”

“Yes.”

I was silent for a few beats.

“Thank you,” I said quietly.

“You are most welcome.”

He pulled his plate back just as I reached for another piece of his bacon. Then something strange happened. We both laughed.

No. Not again. Never again.

I shoved my plate to the center of the table and stood up so quickly, I nearly knocked my chair over. I turned and practically ran from the room.

"Damn it," Draco sighed.

He stood, grabbed his phone, and flicked a rune over his shoulder as he left the kitchen. Shadowy figures emerged from the walls and began collecting dishes to take over to the sink to wash. He unlocked his phone and scrolled through his contacts, stopping to select Morgaina.

"Hello?" She answered breathlessly.

"Morgaina."

"Oh, Draco. I'm so glad you called," she gushed.

He squeezed his eyes shut, already annoyed. He plastered a smile on his face and tried to summon some charm.

"I need your help, little one," he said. He heard her squeak on the other line.

“Anything for you.”

“Good. I need to get in touch with Isaiah.”

Silence.

“Morgaina?”

“Yes,” she said in a flat voice.

“Is there a problem?”

“He’s difficult to find.”

“I know, but you are talented at finding things. It’s important.”

“This is about her, isn’t it?”

“Morgaina,” Draco said, menace lacing his voice, “choose your next words carefully.”

“Fine,” she said, blowing out a breath. “It’ll cost you.”

“I’m aware.”

“Are you?”

Draco didn’t know what to make of that comment, but suspicion crept in.

“What do you have in that little head of yours, fae?”

"You."

He almost threw the phone.

"Out of the question. Don't be a fool."

"A date."

"No."

"A kiss."

"With a fae? Not a chance."

"Fine. Pay me triple my usual fee plus a nice piece of jewelry, and I'll do it."

"Why the extortion, Morgaina?"

She dropped her voice to a whisper. "Things are getting tense around here, Draco. I'm getting out of here at the first chance." She paused. "Is this about Damien being back?"

"How did you know about that?"

"I'm in Faerie. We hear things."

"Fine. Deal. Same meeting place?"

"No, we need to go somewhere else. He has his spies there."

"Deodamnatis," he swore, running his hand over his mouth.

"Meet me at the Hedge near PCH. I don't want to be in your world any longer than I have to."

"I'll be there in an hour."

"See you then."

He hung up and stared at his phone. Faerie was stirred up about Damien. He had reinforcements.

"Shit."

~

When I came out of my room, Draco looked ready to head out.

"Going somewhere?"

"To meet the contact who can get us in touch with an Ancient."

"That was fast."

"Well, as you said at breakfast, time is a factor." He flashed me one of those devilish smiles.

"Do you… want company?" I choked out. I would have laughed at the stunned expression on his face if my heart hadn't been hammering against my ribcage.

"Ordinarily, I wouldn't refuse, but this person is very shy about her identity."

"Her?"

"Yes."

"Okay, then."

I turned to go, trying to ignore the twinge of jealousy squirming in my gut. I'd had boyfriends since Draco and I split up. Why should he be seeing a woman make me feel crazy?

"Angelica." I turned back around and feigned nonchalance.

"Yup."

"She's a contact who provides me with items I can't get anywhere else. She's fae and irritating as hell."

"It's really none of my business."

"No, I suppose it isn't."

He turned on his heel and strode out the front door, muttering to himself. Soon, I heard a car engine roar to life and gravel spray as he sped off.

Draco's mind whirled. He kept getting caught off-guard. The short conversation they'd had as he was leaving and her thinly veiled jealousy had him feeling wretched. He cranked up the music to try and drown his thoughts.

I mean, it isn't as if I'm sneaking off to fuck some woman. I'm trying to help, for pity's sake.

Then he felt another pang of guilt for even having such an angry and defensive thought. It just… stung.

Through his defensiveness and hurt, the thought he'd been trying to ignore poked up, unbidden, into his mind. He had to admit he liked the fact she was jealous.

He wound his way through the canyon toward the spot Morgaina had designated. Soon, he pulled into a small parking lot and walked a short way to an unassuming outbuilding just inside the front gate of Will Rogers National Park. Distracted by his inner dialogue, he didn't see the body until he nearly stumbled over it. His head whipped around, scanning the area. Whatever had done this to her was gone.

Her shirt was burned, and her face was bloodied. Draco picked her up and cradled her, quickly carrying her toward the small outbuilding to take a closer look away from prying eyes. She was so small she felt like a child in his arms. His stomach clenched. She'd been about to take off and disappear before he'd called her. She was only here because he'd asked her to be.

Stop it, Draco. Feeling guilty over this won't help anything. This is Damien's fault, not yours.

He grimaced. Despite the logic, he still felt responsible. He reached the outbuilding and locked the door behind him. Without the fae magic to open the portal to Faerie, he could at least send a message. He gently laid Morgaina down and crouched next to her.

Examining her wounds, it was clear a demon was responsible. Blood surrounded the red on her shirt, and he lifted the fabric to reveal a cauterized wound the size of a grapefruit on her right side. A slightly smaller one marred her left breast. Something had gored her from her liver to her heart, likely killing her instantly. Small mercies.

He rocked back on his heels. Peeking out of her bra was a scrap of paper. He squinted and leaned forward, snatching it with his fingertips. Scribbled on it were "**49:14**" and a phone number. He shoved the slip of paper into his pocket and approached the mirror.

He retrieved the small diamond tennis bracelet he had intended for Morgaina and placed it on the edge of the sink. Peering into the mirror, he pricked his ring finger with his penknife and pressed his hand against the reflective surface. With the message sent, she'd be taken care of from here by those on the other side of the Hedge.

He left the building and made his way back to the car.

~

Kaden and I were sitting at the kitchen table having coffee when Draco returned. Noticing the bloodstains on his clothes, I felt a stab of fear. His face looked drawn and weary. I put my coffee down and stood up, almost taking a step toward him before stopping myself.

"Is everything okay?" I asked, focusing on keeping my voice steady.

"No."

My heart lurched.

"What happened?" This time, I did take a step closer.

"A demon slaughtered Morgaina just outside the Hedge. The portal looked untouched, but I think Damien has more reinforcements than we originally feared."

"That's not good news," Kaden remarked.

"What was a demon doing that close to Faerie?" I felt sick.

Typically, demons left the Courts alone. Bringing a fight to them was akin to suicide.

“Morgaina mentioned on the phone that Faerie is getting nervous about Damien’s return. She planned to take the money she got from me today to leave.” He ran his hands through his hair. “I did retrieve this.” He showed us a small piece of paper with a phone number and a Bible verse.

“I need to get a Bible from the library,” he said, turning to leave the kitchen.

“Isaiah **49:14**: ‘But Zion said, The Lord has forsaken me, the Lord has forgotten me,’” Kaden read from his phone. Draco looked at him, and he shrugged. “In Google, we trust, man.”

“Well, that’s cryptic,” I said, sitting back down.

“Isaiah has been off the map for a long time.” Draco tucked the paper back in his pocket. “I just hope this isn’t literal. If he’s disconnected, I don’t know who else to turn to with enough knowledge and power to raise golems.”

“It’s worth a try. Give him a call.”

“I will. I need a shower first.”

What’s going on? Why is there blood on you, friend?

"Had a little run-in with the murdered body of Morgaina."

That's unfortunate. From whom are you to procure phoenix blood now?

"Phoenix blood?" I asked, suddenly suspicious.

There were very few spells which required that ingredient and none of them gave me the warm fuzzies. Draco glanced at Kaden.

"What were you using the phoenix blood for, Draco?" I felt my face heat with mortified anger. Now, I could think of the one spell which would account for the guilty look he was giving me. "You cast a voyeur spell?" I asked quietly.

He had to keep an eye on you, Angelica. Otherwise, he would have lost his mind.

"I'm sorry, Angelica. I know how this looks—"

"I don't think you do." I turned to Kaden. "Did you know about this?"

"He told me last night."

I made a disgusted noise. "How long, Draco?"

His shoulders slumped. "Since a couple of months after you left."

I turned and left, slamming the door behind me.

"That went well." Kaden collected the coffee mugs and brought them to the sink. Draco tilted his head, listening.

"That's not how it works, Mischief."

"I really wish I could hear her side of the conversation," Kaden said, rinsing the mugs.

Mischief padded over to Kaden and rubbed against his leg. *My apologies. We were being rude. Better?*

"Whoa. You are one powerful little familiar, and you know that?"

Practice. I was just wondering why she was so upset. I would have thought she'd be flattered.

"Maybe deep down, Mischief. Right now, she's pissed."

But why?

"Because it's a massive invasion of privacy, and I doubt Pepper would have been comfortable knowing the man who betrayed her was watching," Kaden responded.

"Thanks for that," Draco said, and Kaden nodded. "For what it's worth, watching her suffer was a special type of torture for me." He paused. "And I know a little about torture."

"She will definitely not see it that way. And you've mentioned knowing about torture before. What's the story there?"

"Hm. Let's just say at one time I belonged to a coven with some old-school values, and I had a specific talent they appreciated."

Kaden took a half-step away from Draco.

"Wow. That's… wow."

Draco nodded with a grim expression.

"I'll be in the shower if anyone needs me."

Kaden nodded. Soon, he heard the upstairs bathroom door shut.

Should I go to Angelica? Mischief asked.

"You might want to leave her alone for a bit," Kaden replied, stroking her head.

Why?

"Because she's angry right now, and even though it may not make sense, she might end up taking it out on you."

I think I understand. She's like a wounded animal.

"Perfect analogy. And you're more right than you know."

~

Damien's house in the Midwest remained exactly as he'd left it. The two girls were still confined in the basement, now starving and dangerously dehydrated. He strode in, and the girl on the bed started trembling with fear. The other didn't react at all. Unconscious. He had brought water for both of them, but after assessing the state of the one chained to the wall, he decided on different plans for her.

He drew closer to examine her. Still alive. Good. Sores had formed on her wrists where the shackles had rubbed the skin raw. Damien inhaled deeply, savoring the scent of decay setting in. He lightly tapped his palm with the flat of the blade, deliberating his next game. So many ideas. Such fun to be had.

Faeries And Demons And Golems, Oh My

“It’s similar to the containment spell you used last time but with a little extra power.” Draco had a large tome open on his lap while Kaden leafed through an ornate Bible.

“Extra power? How so?”

“This time with me.”

“Right. Of course.”

“Do you think the verse is some kind of code? Or possibly a password?” Kaden called from the corner of the library.

“Well, I tried the phone number, and it didn't work.”

A thought occurred to me. “Maybe it’s coordinates.”

They both looked up at me.

"That's brilliant." Draco jogged to the massive map on his desk.

"That would place him somewhere in Canada. Near B.C."

"I think you've cracked it, Pepper." Kaden grinned at me. I grinned back.

"Okay, assuming we have the coordinates for his location, let's work on these spells before we take a little road trip."

We spent the rest of the day committing the runes and power words to memory. Seeing how well we worked together, hope flared in my chest again. This time, I didn't quash it so quickly. We had a plan.

The next morning, we packed for our trip to Canada. Draco insisted on taking his Range Rover, arguing it had the room for our gear and the capability to handle rough terrain if necessary. I voted for the Bentley, and Kaden wanted the Maserati, but we had to concede to logic. Hopefully, there will be time later to play with the other cars.

"I'm all packed. Need any help in here?" Draco leaned against the doorway of my room, his arms folded over his chest.

"I think I've got it. Have you seen my cards?"

Draco slowly entered the room as if expecting me to throw him out at any second. His eyes roved the space, lighting on a pile on the desk. He walked over and lifted a book, revealing my deck beneath.

"May I touch them?"

I nodded mutely. He reverently cradled my Tarot cards in one hand, running his thumb along the edges.

"I… haven't had a reading since you've been gone. Would you do mine now?"

I shook my head. "I'm not sure that's such a good idea."

"Why not?"

"I'm a little… distracted. The space I'm in is not conducive to an accurate read."

"Try me."

I sighed. "Fine. Shuffle."

He cleared a space on the desk and began shuffling slowly. I watched his forearms ripple as his hands danced lightly over the cards. Walking over, I settled into the chair as he cut the deck three times.

"How thorough a read would you like? One, three, five, ten, or twenty?" I asked.

"Three should be sufficient," he replied.

"Okay." I fanned the cards, allowing him to pick his three cards before I selected the significator.

"Significator, Magician." My heart jumped. "They are all Cups, meaning your emotions are in the driver's seat right now." I cleared my throat.

"First card, five of Cups. You possessed love and bounty but did not appreciate what you had and were dissatisfied. No shock there," I added with a roll of my eyes. "Second card, eight of Cups. You have spent a long, lonely time mourning an emotional loss. Again, no shock." I pointed to the final card and stood. "Third card, two of Cups." I closed my eyes, fighting tears. "Friendship, union, and reunion, lo-" My voice tripped on the lump in my throat before I could finish the word 'love.' I glanced up at him, and judging by the tender look on his face, I might as well have said it. I should never have taught him anything about the stupid Tarot. I cleared my throat again but still couldn't seem to find any words. I picked up my deck and started shuffling.

"Why is it referred to as having a frog in your throat?" I blurted out, my mind racing ahead of my filter.

Wait, *what*?

He blinked at me. "Um…"

"Because between the frogs and the moths, I have the regular menagerie going on." I mentally scolded myself. Stop talking, Angelica.

"Moths?"

"The convention." That conversation was with Kaden, not Draco.

I actually knew what I was talking about and thought I sounded unhinged. I set my deck down and screwed my eyes shut. Ostrich mode: Engaged. I couldn't see him, therefore, I was invisible.

I cracked one eyelid open to peek at him and saw his shoulders shaking with silent laughter. I sat back in the chair and dropped my hands to my lap.

"Fine, fine. Yuk it up, chuckles."

He snorted, his hand flying to his mouth and his eyes widening. I grinned.

"Ha!" I pointed at him and started laughing.

Soon, we were in full hysterics, tears streaming down our faces, and neither of us could remember what got us started. I threw my head back, succumbing to another peal of laughter, and the chair tipped. It teetered on its back legs, apparently deliberating over whether I had filled my daily mortification quota before judging my metrics sub-par.

I watched the room twirl and tip around me in slow motion, just catching sight of Draco trying to catch me before I crashed to the floor. He overshot, I caught him in the knee with a flailing limb, and he ended up falling right alongside me, actually*, on top of me.*

We froze and stared at each other in shock, panting from the laughter and winded from the fall. Neither of us was laughing now. I was consumed by his scent, in his presence, his eyes… His eyes had always been my favorite feature of his. Hazel - mostly green but flecked with gold and copper, the irises ringed in milk chocolate. I'd spent so much time studying them which I knew them better than my own.

Those gorgeous eyes filled with heat, and my heart beat faster in my chest. The room around us disappeared, and my awareness zeroed in on every place my body touched him. It wasn't until his gaze dropped to my lips that my excitement turned to panic.

He was too close. We were too close. Too much. Too soon. Not now. Not ever. Never again.

Never. Again.

I started to sit up, and he jumped off of me like a scalded dog. I pulled my knees to my chest and wrapped my arms around them. Pressure built behind my eyes, and my throat started to close up. I was too close to tears to chance looking at him. I couldn't look at him even when I asked him to leave.

I waited for him to shut the door before I allowed the first tear to fall.

~

When the girl's heart stopped, Damien set the device down and licked his tongue over her face. He savored the intoxicating mixture of blood and tears, surrounded by the scent of death. The girl behind him wept. He tilted his head back, allowing his true form to manifest.

Turning his obsidian gaze to the terrified girl, he watched her eyes widen at his grotesque appearance. He knew she would be unable to look away now that he had locked his eyes with hers. Spreading his arms wide, he reveled in her delicious horror before bursting into booming laughter as she quickly lost her mind.

~

I chose to sit in the back, giving Kaden a shotgun. I watched the scenery roll by, transfixed by the ocean on our left and mulling over the most recent turn of events. For the second time in a week Kaden and I were traveling PCH, this time headed North with Draco in the car. We were listening to music with the windows down, your typical California youth on a trip up the coast. The day was perfect and clear, Catalina visible on the horizon.

I had missed the Pacific. I'd missed the golden light dappling the water and the life surrounding it, animal and human. I'd even missed the horrid seagulls circling and cawing above, bullying their way into people's lives and food.

"Days like this make me almost understand why you live here, Draco." Kaden leaned his head back and closed his eyes. "This place would be perfect without the people."

But that's the best part of the scenery, new friend.

He chuckled and reached back to scratch Mischief behind the ears.

“We’ll make a convert of you yet, K.” Draco signaled, passing a wide-eyed tourist going five miles under the speed limit. All three of us turned to look at the middle-aged woman with sunburned shoulders in a sleeveless top.

“You guys hungry yet?”

“What did you have in mind?”

Mischief purred next to me.

“Well, there are snacks packed, but we can stop if you guys feel like something more substantial.” Kaden hummed his assent. “How does Geoffrey’s sound?”

“I could slam back some seafood. But don’t we need a reservation?” I asked, fidgeting with my deck.

“I don’t.” Draco caught my eye in the rearview and winked. I responded with a small smile.

“What’s Geoffrey’s?” Kaden asked, rolling down his window.

“It’s a seafood joint up the coast a bit. I think Draco wants to wine and dine you, K.”

"I'm a willing escort if good seafood is involved, Pepper."

"You should probably lay down some ground rules then."

"I dunno. I'm pretty morally flexible when it comes to seafood."

"I was talking to Draco." I grinned at Kaden and he blew me a kiss. Draco laughed and took a swig from his water bottle.

We rode in companionable silence for a while, watching the California coastline fly by. It was beginning to feel as if we were all falling back into our old rhythm with no time passed. I had to admit it felt good. We had a powerful ally against Damien and Kaden didn't seem as if he wanted to kill Draco anymore.

I wasn't ready to be honest with myself about the other reason the easy camaraderie was making my heart flutter.

"Does Geoffrey's have a dress code?" I furrowed my brow. "We may need to stop and pick up a few things if they're going to let us in."

"Don't worry about it. We're going to be there before the doors open. We'll dine at the Chef's table."

Kaden whistled. “Fancy, indeed.” Draco gave him a lopsided grin.

“I helped Bijan out with a problem some years back. He’s always quick to show his appreciation whenever he sees me.”

“Bijan?”

“Yes. Bijan Shokatfard. He’s Chef there.”

Even I couldn’t hide how impressed I was.

“You know, if this were a date, you’d definitely be scoring points.”

“That never even occurred to me.”

“To take a date there? How could it not?”

He was quiet for a beat.

“To go on a date.”

“Oh.”

Kaden gave me a look in the rearview, and I sat back in my seat. Fortunately, we arrived at the restaurant before the silence turned awkward.

True to his word, Draco was treated like royalty when we showed up in our jeans and tennis shoes. They ushered us back to the kitchen and Bijan came around to shake hands with all of us, thumping Draco heartily on the back. He presented a couple of creations he'd been working on and served us himself. Must have been some favor.

After lunch, our trip continued without incident, and we were approaching the Redwoods when we decided to stop for the night. After a short debate, we decided on a versus camping. Mostly due to the fact that it is easier to find a hotel than to find a legal place to camp in the area. Plus, I was looking forward to a long, hot bubble bath after spending the entire day in the car.

Draco secured us a double en suite so we wouldn't be separated. I slid my doors closed and ran the water, emptying the entire complementary bubble bath gel in the water as it filled, and soon the room was steamy with a delicate scent of sugared violets. I undressed and slipped in, sinking all the way to my chin with a sigh.

We would reach Oregon tomorrow and likely Washington before the day's end. I closed my eyes and allowed the tension I'd been holding to be carried away by the bubbles. I pictured the ocean as I had seen it earlier, sunlight skipping over the waves and birds gliding on the wind. A smile crept on my face.

Just as I opened my eyes to reach for the loofah, an invisible force pinned my arms to my sides and shoved my head under the water. The sounds of my struggles were muted beneath the surface. I thrashed against the weight to no avail. Soon my lungs began burning. I felt myself weakening and slowed my movements to conserve what oxygen I had left.

I tried to conjure runes to counteract whatever spell was holding me down, but without being able to properly move my hands, they fell flat. I began to see stars and my limbs grew heavy. Darkness crept into my peripheral vision. I was dying.

Then I shot forward, gasping for air. Kaden and Draco flanked the tub, casting and chanting. They stopped once they realized I was breathing, and Kaden spun away from my nakedness. Draco stepped forward and cupped my face in his hands.

"God, Gelli. Are you alright?"

He knelt down and I threw my arms around his neck, gulping lungfuls of air.

"It's okay, sweetheart. I've got you. It's okay."

He stroked my hair, and I gripped his shirt, sobbing against his shoulder.

“I’m going to finish warding the rooms.” Kaden started to leave, but I stopped him.

“Kaden. Thank you,” I said in a hiccuping voice.

“Of course,” he said over his shoulder.

He left, shutting the door behind him, and I was suddenly aware of my nakedness. Sensing my discomfort, Draco let go of me and retrieved a towel and a big, fluffy robe. I nodded my thanks, and he turned around to allow me some privacy to towel off and robe myself.

“And thank you, too, Draco.” He turned back around and gave me a small smile.

“No need to thank me, Gelli.” He furrowed his brows. “I think it's best I stay with you tonight. I’ll stand watch while you sleep.”

“You really don’t have to do that.” I pulled the stopper on the tub.

“I insist.”

I paused. My hands were still shaking and buzzing from the adrenaline.

“Okay,” I relented, and he nodded in relief.

“Why don’t you get into bed? I’ll take the couch.”

“Um, Draco?”

“Yes.”

“Could you…” I couldn’t get the question out. How could I possibly ask him to stay in the bed and hold me?

“Angelica,” he said, his expression soft. “I’ll stay wherever you want me to.”

I blinked back the tears which sprang to my eyes. He held out his hand, led me to the bedroom, pulled back the covers, and indicated I should crawl in. I complied, and he tucked the covers to my chin before laying beside me atop the blankets.

“I’ll be right here, okay?” He murmured. I nodded with a sleepy smile.

“Thank you.”

“Anytime.”

I turned on my side and was on the brink of unconsciousness when I felt his arm tentatively snake around my waist. I tensed slightly before relaxing against him. Soon I was fast asleep.

The smell of breakfast woke me the next morning. I stretched and yawned, disturbing a sleeping Mischief on the pillow next to mine.

Good morning, Angelica.

“Good morning, sweetie. What are you doing here?”

Sleeping.

“Well, I get that. I meant, what are you doing in my bed?”

I sleep where my friend sleeps.

I froze mid-stretch as the previous night came back to me. The bath, the attempt on my life, the boys breaking the spell, Draco tucking me in. Falling asleep in his arms.

I ran my hand over my face and got up. I was still wearing the bathrobe from last night, so I dressed quickly and followed the voices into the dining area.

Just like at Draco’s, the table was laden with a feast of breakfast fare. A small moan of appreciation escaped my lips before I could stop it. Both boys looked up with expressions of surprise.

"What? It's breakfast. Do I seriously have to stand on ceremony around you two now?"

"It's not that, Pepper. It's… well, it's your hair."

"What about my hair?"

"Just go look."

Suddenly nervous, I strode quickly to the bathroom mirror and dropped my jaw at my reflection. My curls were a little wild from having slept with wet hair, but what drew my eyes was the color. Instead of the strawberry-blonde I was expecting, my hair was now a deep, glossy chestnut with reddish highlights. Even more alarming was the shock of white framing my face.

I lifted my hand to touch the locks around my cheeks and the Angelica in the mirror did the same. It was really me. Holy shit. Clearly whatever spell I had been hit with last night was a damned heavy one.

"What the fuck??" I exclaimed, slamming a fist down on the bathroom counter.

"You okay in there, Angelica?" Draco called, then winced at the stupid question. "No, I'm damned well not okay!" I marched into the dining area and fell into a chair. "I mean, what the hell?" I turned to Draco. "Have you seen anything like this before?"

He nodded. "A couple of times at my old coven. Some of those witches were into some heavy, black magic. Really nasty stuff."

"Not helping."

He responded with a grim smile. "I'm sure it comes as no surprise to you, Angelica. The spell you were hit with last night was not cast with good intent."

"I know. You're right," I huffed out, slumping in my chair. "It's just scary as hell." Draco nodded again.

"I think the most important thing to assess right now is whether you have any residual negative magicks clinging to you, Pepper," Kaden said, turning to Draco. "We need to do an MRI."

Kaden had coined the term "MRI" for an aura and magic scan, inspired by the medical procedure of magnetic resonance imaging. His version stood for "Magic Residue Inference." It was a completely fabricated name, but an accurate description of the process, nonetheless.

I agreed. If there was something nasty clinging to me as a result of being ambushed by a black magic or demon spell, we needed to know what it was in order to counter it. We decided to start right away.

Kaden left shortly after breakfast to collect a few things for the spell, which left me and Draco alone. I watched him agitatedly pace the rooms for a good half-hour before calling to him to stop.

"Maybe a little cribbage?" I suggested. Draco was a killer cribbage player, and we'd always had fun trading wins back and forth. He perked up.

"You have a board with you?"

"I do."

"I was thinking about asking for another read - a longer one - but cribbage sounds like a lot more fun."

I smiled at Draco and retrieved the board. He pulled a Jack and dealt first. We were in a dead heat at 5-5 when Kaden returned. Draco was in danger of being skunked, and my current hand was going to rocket me to a win. When Kaden blew in, I simply showed Draco my cards, and he shook his head, laying his down.

“The victor and Queen of all things cribbage,” I declared, slowly raising my fists above my head in a victory pose.

“Alright, go shower, Your Highness.” Kaden set his supplies on the table in the dining area.

“Yes, Sir K, Knight of the Cribbage Court.” I stood from the table and stretched my back. “Loser gets to clean up.” I gave Draco a smug smile.

“I had you right where I wanted you. And hold it—” He sat back.

“I don’t get a title in Her Royal Highness Angelica’s Court of Cribbage?”

“I hereby appoint you to the position of Her Royal Highness’ Cribbage Court Jester.” I picked up the cribbage board and tapped each of his shoulders in turn.

“You honor me, my Queen.” He bowed low, took my hand, and brushed a kiss along my knuckles. An awkward silence followed.

I cleared my throat. “Off to the shower with me, then.”

“Use the rose water!” Kaden called to my retreating form. I waved over my shoulder in acknowledgment.

The boys prepped the room while I showered. Preparation for these readings required precise attention to detail. I meditated while bathing with rose water, selenite, and onyx crystals strategically placed around the tub. An amethyst geode lay over my heart chakra as I concentrated on purifying my essence in both spirit and thought.

Thinking of nothing is more challenging than one might assume. Despite my well-practiced meditation techniques, it took over twenty minutes to feel centered enough to wash my hair. Soon, I was toweling dry with the blessed shroud Kaden had set on the counter.

Draping a second swath of purified fabric around my body, I padded barefoot to the circle the boys had set up while I showered. Kaden and Draco stood inside the ring of salt and herbs. Approaching, I asked the Goddess for permission to enter. I genuflected before the candles at each of the four directions, beginning with North and moving deosil around the circle. Mischief sat on a makeshift altar, her tail wrapped around her paws. Kaden broke the circle with his athame to allow me entry, and I lay at its center, my body like a compass with my crown chakra pointed North.

We were ready to begin. Kaden and Draco closed the circle, moving widdershins and then deosil, before kneeling and gently lifting my arms. Kaden took my right, and Draco my left. Each bent to place his forehead to the inside of my wrists, chanting in low tones.

My body pulsed with power. Each of them was a strong mage in his own right, and their combined energy was electrifying. I breathed through the surge of energy lighting my insides on fire.

"So we begin," Kaden said, and they got to work.

I lost sense of time as they brushed their hands inches above my skin. It felt like a long while before Draco finally spoke.

"I've got something."

My heart jumped. So, something had been left behind by the spell other than my new hairstyle.

"Are you able to get it, or do you need help?" Kaden asked, still running his hands along my aura.

"I'm going to try on my own first. I'll let you know if I need you to piggyback."

Draco began chanting again, and I started to feel a tug on my chest. My breathing became shallow and labored as pain bloomed in my ribs. A sheen of sweat glistened on his brow.

"It's stuck. I need you, K."

I opened my eyes to see tendrils of black snaking from my heart, winding around Draco's hands as he pulled. Kaden stepped behind Draco, placing both hands on his back and closing his eyes. Another surge of power lit me up, and I could feel the two mages' magicks intertwined, racing through my body. With one last pull, I gasped, feeling a rib snap as Draco yanked, stumbling back into Kaden.

"Break the piggyback! Quickly!" Draco shouted. The candles sputtered in a howling wind.

Kaden dropped his hands and intoned a spell for cleansing and protection. Draco wrestled the entity into a small binding vessel, plunging his hands into the copper jar and quickly pulling them free. He slammed the lid closed and fell back onto the carpet. He chanted the same cleansing and protection spell as Kaden and let out a long breath.

I concentrated on not passing out from the pain in my chest as the boys moved widdershins, opening and releasing the circle. Mischief hopped down from the altar and wobbled over to me. She slunk up to my side and placed her paws on my chest while the boys grounded. Tears of pain rolled from my eyes as I felt the rib knit itself back together. Broken ribs really hurt like a bitch.

“Thank you, Mischief,” I whispered once I could take a pain-free breath.

You’re welcome, Angelica.

“What would we do without you?”

Go into battle with a broken rib.

I laughed softly. “You’re right about that.”

Most things.

“That was a nasty one,” Draco said, still breathless.

“It was, indeed. Good thing we got it out when we did.” Kaden paused. “I wonder what its intent was.”

“Likely a tracker.” Seeing the beginnings of panic in my expression, Draco continued. “Gelli, we don’t know it was one of Damien’s. Besides, even if it were, I’d be surprised if he didn’t already know you were on the West Coast, seeing as you were attacked in the hotel. Now, he won’t be able to follow your movements further.” Draco addressed us both. “There’s no way he could know we’re headed to Isaiah.”

Kaden's brow furrowed. "If it was one of his, his spies are getting more subtle. I didn't detect anything when we checked in, and I'd be lying if I said that didn't scare me."

"Me, too," I croaked out. "God, I'm starving."

"That's to be expected." Kaden smiled at me.

"We set aside some extra breakfast food. Give me a minute, and I'll heat it up for you."

"You're the best."

"I know."

Kaden walked over to Draco and helped him off the ground before making his way to the kitchenette. Draco came over and crouched beside me.

"Mischief took care of the rib for you?"

"Yes."

"I'm sorry about that. It was the only way to get that thing out of you."

"I know. I don't blame you. I'm just grateful she has grown so strong in the healing arts."

Thank you, Angelica.

"You are quite the amazing familiar."

Draco turned and fondly stroked Mischief. She purred and bumped against his hand in an uncharacteristic display of affection.

You guys flatter me. I'm just doing what I was meant to do.

"Well, I appreciate it anyway, little one," I replied, joining him in petting her. She purred louder.

I can't remember...

"Can't remember what, sweetie?"

I can't remember the last time I had both of your attention at the same time. It's nice.

It had felt so natural, it hadn't even occurred to me how intimate a moment like this was. I gave her one final scritch behind an ear and dropped my hand. She stretched and trotted off.

"Well," I said with a groan, "Shower number two for me. I need to get all this anointing oil off."

"I'll help Kaden in the kitchen."

Draco stood up and extended his hand to help me. Once I was standing, he didn't let go right away. I fought the urge to close the distance between us, sensing he was struggling with the same idea. His hand was warm and smooth, feeling perfect in mine. Energy pulsed up my arm, and goosebumps rose on my skin. I glanced at Draco, catching him looking at our hands with a mixture of trepidation and awe. Finally, I gave a gentle tug, and he reluctantly dropped my hand. I turned toward the bathroom but paused at the door.

"Oh, and Draco?"

"Yes?"

"Thanks again. It seems you're making a habit of saving my life." He nodded with an enigmatic smile, making me briefly think how I'd like him to join me. *Not gonna happen.* I walked into the bathroom and shut the door.

Kaden and I sat in the Range Rover with our packed bags while Draco settled the bill. Kaden twisted around in his seat to face me.

"I'm sorry I was unable to detect anything unusual in that hotel, Pepper."

"Not your fault. Besides, you more than made up for it."

"You could have died."

"But I didn't."

He chewed his lip in thought.

"If Draco hadn't been there, you would have." His voice was so low I had to strain to hear him.

"What are you talking about, K?"

"It was too strong for me. Without him, you would have drowned in front of me."

There was a terrifying thought. Kaden was an impressive mage.

"Then I guess it wasn't such a bad idea to involve him after all," I said softly.

We fell into a contemplative silence, and I shuffled my deck. The weather was cool and moist, and the air smelled of damp earth and growing things.

The driver's side door opened, and Draco entered the car, bringing the fresh scent of the outdoors with him. It complemented his own scent perfectly and I fought the urge to close my eyes and breathe deeply. To let it surround and fill me. My inner voice nagged me, suggesting it wasn't the only thing I wanted to surround and fill me. I told it to shut its cakehole.

"You guys ready to get going?" Draco asked.

I ducked my head to hide the blush creeping up my neck. I'd blushed more in the last week than I had in years. Damn him. I hated blushing.

"As ready as I'll ever be," I replied, shuffling the Knight of Wands back into my deck.

Having pulled the card just as he asked the question, it was clear the answer was yes, and we were embarking on an adventure. Whoopee.

"Think we could stop for some coffee?" Kaden said around a yawn.

"Of course. Is drive-through okay, or do we need to actually go in?"

"Drive-through is perfect. I don't need the ambiance, just the caffeine."

"I second that," I said, tucking my deck into my coat pocket.

"Mischief, you want to get in on this too? Espresso?" Draco joked.

I don't like coffee. I'll take steamed milk, though.

"Alright. Let's go. I'll treat," Draco said, firing up the engine.

"I'm not going to turn that down," Kaden said with a smile. He leaned his head back and closed his eyes.

That was when the truck hit us. The force of the collision sent our SUV careening, tossing us like dice on a craps table. I barely managed to cast a shield around myself and Mischief before we slammed against the door, snapping my collarbone on impact. I cradled Mischief in the crook of my good arm while she trembled with terror.

I blinked blood from my eyes, trying to see into the front seat. From where I lay on the floor, I could see Kaden bleeding freely from a wound on his forehead, shaking his head as if to clear it.

"Draco?" I gasped, holding Mischief closer. "Kaden? Are you guys alright? Draco, I can't see you from here." Kaden didn't answer.

My friend is hurt. He's hurt really badly.

Mischief squirmed in my arms, trying to get to the front seat.

"Okay, hold on, little one. Let me get you up there."

I tried to hoist myself onto the seat and collapsed with a cry. My shoulder was on fire.

"Sorry about this, but it's the only way."

I scruffed Mischief and shoved her through to the front armrest. She took a moment to get her footing before darting to the driver's seat. I still couldn't see Draco.

"Kaden! Answer me, dammit!"

"Huh?" he finally responded, and relief flooded through me.

"Kaden, please check Draco! I can't see him!"

"Right…" He swung his head to the left. "He's not moving, but the kitty is kneading her paws on his chest. Little wobbly kitty muffins. How cute…" He began to close his eyes.

"No, Kaden! Stay with me!" I tried to get up, and frustration welled when I again crumpled back to the floor with a gasp.

Angelica! He's not moving. Please, I need your help, please…

"Oh, God…" I moaned. I gritted my teeth and pulled myself up to the partition between the front seats, crying out when my shoulder brushed against Kaden's. I looked over to Draco and saw Mischief intently kneading her paws on his chest.

I'm trying, Angelica. Please, please help.

"Okay, Mischief. I'm here."

She shoved her head beneath his chin with a desperate sound. I reached over and felt his neck for a pulse. I couldn't find it right away, and my heart plummeted. I dug around, blinking away blood and tears until I finally caught the faintest of beats. It was shallow and erratic. He was dying.

"Mischief, what do I do? Walk me through it!"

He's stopped breathing, Angelica. Please help!

"I'm here, Mischief. I'm here. Tell me what to do."

Place one hand over his heart and another on his side.

"We're going to defibrillate him?"

Yes. And now.

I followed her instructions, placing one hand over his heart and another on his side. Despite the pain in my shoulder, I threw everything and focused all my power into my hands. His body felt unnaturally still. I choked back a sob and intensified my efforts. My hands began to warm.

His heart's stopped, mama! You have to do this now!

Blocking out everything else—the pain in my head and shoulder, Kaden's incoherent mumbling, the crackling flames around the car. All I could see and feel was the power in my hands. They began to burn. Clenching my jaw, I poured all my energy into him. His back arched, hands twitching, but he still wasn't breathing.

"Drake, you have to come back—" I swallowed hard. "Please don't leave me..." I whispered near his ear.

You have to do it again.

I repeated the process. I ignored the voice telling me I had nothing left and pushed through. He arched upward and jerked again. This time, he let out a long moan, which ended in a cough. Blood dribbled from the corner of his mouth, and his eyes remained closed.

"Please, Drake. Please talk to me."

"Gelli…"

I choked out a relieved sob.

You did it, mama. Oh, thank Bastet. And thank you. Thank you so much.

"What happened?" Draco slurred.

"You were—" Movement in my periphery caught my eye. The driver of the truck was getting out. A very large man hobbled his way toward us. Then he was close enough I could see his black eyes.

Demon.

"Oh, fuck. Guys! We gotta go! Now!" Neither of them responded. Kaden seemed to have passed out again, and Draco had closed his eyes, breathing deeply. "Mischief, I need your help!"

How can I assist?

"I need you to heal my collarbone right now. This guy is coming for us, and the boys are down for the count."

Grab me.

I lifted her up and leaned back slightly so she nestled closer to my chest. She placed a paw on each side of the break, and immediately, I felt the hot and cold magic as she began healing. The demon was drawing closer. I bit back a cry as the bones in my chest shifted, knitting back together.

As Mischief worked, I weaved a protection spell around the boys to keep them safe while they were unable to defend themselves.

Almost there.

With a pop, the pain ebbed, and strength returned to my arm.

"Great job, Mischief. Thank you. Now, focus on the boys while I take care of this other problem."

I placed Mischief on the armrest again and began crafting battle runes. My door was crushed shut, so I crawled through the broken window. I ignored the glass cutting into my hands and jumped to the ground to face the demon.

"C'mon, you motherfucker."

As soon as he came within range, I hit him with a paralysis spell, quickly followed by a ballistic rune. He staggered back but didn't fall. I swore under my breath. After expending so much power into Draco, my tank was running low. I hit him with the same combination, yet he remained on his feet. Clenching his fists, he advanced, dragging one leg behind him. He unhinged his jaw and screeched in a way reminiscent of hellhounds. I screamed back and shot a fire rune directly into his throat. He exhaled, sending the flames back at me close enough I caught the putrid scent of rot and brimstone.

Then he was on me. He backhanded me and I stumbled back, my ears ringing. I kicked out, trying for his knees, but it missed. He followed up with a punch to my gut before tackling me to the ground. I wheezed and desperately tried to draw air into my uncooperative lungs. He straddled me and reared back to punch me in the face when suddenly he froze and was hurled into a wall, cracking it on impact.

I looked up to see Draco with his arms out and his face a mask of fury. Finally, I managed to draw a painful breath into my burning lungs. Kaden staggered out of the car and strode up to stand next to Draco. He threw his arm in a rough arc, causing the demon's head to spin on its neck.

I let out a breath and collapsed in relief, coughing into the asphalt.

"You okay, Gelli? Are you hurt?" Draco appeared at my side in a blink.

"I'm fine. He just rang my bell a bit." I coughed again, and my head throbbed. Draco cradled my cheek to get a better look at where the demon had hit me. I could feel my eyes beginning to swell shut.

"We should let Mischief take a look at this. Come on."

"No, let her rest. She saved our lives. She must be exhausted." Draco helped me to my feet, keeping one arm around my shoulders. I leaned into him. "Is Kaden okay?"

"He's fine, thanks to you and Mischief." I began trembling as the adrenaline wore off. "Hey… It's okay, Gelli. It's over." I shook my head.

"No, Draco. It's just the beginning. We were caught off-guard, and that's a mistake we can't afford." I stopped and turned to look at him with my good eye. "We need to get to Isaiah. Then we need to get somewhere with a lot of weapons." I sighed. "And we need reinforcements, like yesterday. One demon should not have been able to put us down so easily." Suddenly, I was so tired I could barely keep my eyes open. "But first… I think I need a nap." All of the energy was draining out of my body and the pain was beginning to creep in. I lifted my hand to my side and it came away bloody.

"Jesus, Angelica. You're hurt!"

"No, I'm fine." My voice sounded far away and my vision was tunneling. My knees gave out and Draco grabbed me to keep me from crumpling to the ground. He gently laid me down.

"Kaden! Get Mischief!"

"No… Let her rest. She did well…"

"Don't talk, Gelli."

I gasped as Draco pressed his hand firmly to my side to staunch the bleeding before whipping off his shirt and pressing it against the wound. I just needed to close my eyes for a minute.

"No, Angelica! Keep your eyes open! Stay with me!"

The last thing I noticed was how good he smelled before I passed out.

I awoke in my hotel bed, head and side throbbing. I could only see through one eye. The other was swollen shut. I tried sitting up. Gentle hands pushed me back down.

"Don't move, Gelli. Mischief passed out before she could heal you. She's been asleep for twelve hours."

"Is she okay? Oh, damn…" It occurred to me I had been asleep for twelve hours as well. "We need to get moving. We need to get to Isaiah." I tried sitting up again, groaning at the pain in my side. He pushed me back down again.

"She's fine. Just worn out. And we're safe here for now. Kaden and I both put up the strongest wards we know. We'll get to Isaiah soon enough. He's not going anywhere." He stroked my hair before tucking a lock behind my ear. "Your only job right now is to rest until Mischief is ready again to heal you. And to eat as much as you can stomach." He gave me a small smile, and I couldn't help but return.

As if on cue, Kaden came through the door with a tray piled high with food. He set it down on the table next to my bed and helped Draco get me to a sitting position, propping pillows behind me before laying the tray in my lap. It smelled heavenly and my mouth began watering.

"How are you feeling, Pepper?"

"Like I got hit by a truck."

"Too close to the truth to be funny." He poured me a coffee. "Fortunately, there was no poison on that demon claw or we would have had to deal with more than just stopping the blood."

"Claw? When did he get me with that? All I remember is him hitting me."

"He must have gotten you at the same time. Eat up." I took a bite of bacon and moaned.

"So… good."

Before I knew it, I had polished off my plate, and Kaden was already replacing it with another. It felt like I hadn't eaten in days.

"Boy," I said around a mouthful of eggs, "a fight with a demon sure makes a girl hungry." I shoved a piece of sourdough toast in my mouth. Draco chuckled.

"Easy there, baby. You haven't eaten in almost a day. We don't need you throwing it all up."

I nodded, then froze.

"Baby?" I asked softly. His face paled.

"I'm sorry. It just slipped out." He put down his fork. My appetite vanished.

"No, please don't stop eating," he insisted.

He looked miserable. I slowly picked up a knife and began spreading raspberry jam on my toast.

"It's okay."

"Anyway…" Kaden cut into the awkward silence, "We'll be safe here for now. Just rest until Mischief is ready to work her magic, Pepper."

"Okay." Finally feeling full, I sighed contentedly. "I think I'll try to get a bit more sleep."

"That's what I want to hear," Kaden said with a smile, helping me lean back onto my uninjured side. He placed his hand on my forehead, and my eyelids grew heavy. “Sleep now, Pep.” I let Kaden’s spell wash over me, plunging into a deep sleep.

“We need to talk.” Kaden shot Draco a pointed look and left the bedroom. Draco dropped his head like a scolded puppy and followed. In the den, Kaden whirled on Draco.

“What the hell was that?”

“Like I said, it just slipped out.”

“I told you, you’re not getting anywhere near her. I won’t let you hurt her again.” Draco felt a stab of annoyance.

“Look, Kaden. I have no interest in hurting her. I love her.”

“I don’t care! You gave up that privilege when you broke her heart the first time! You selfish bastard!”

Draco slumped onto the couch. “Don’t you think I know that, Kaden?” He shook his head. “I know I have no right to her. I never assumed that.” He scrubbed his face with his hands. “But I can’t help how I feel.”

Kaden huffed out an exasperated breath and stormed to his room, slamming the door shut. Draco reached up and stroked Mischief, asleep on the back of the couch. She purred softly and twitched her whiskers in her sleep.

I awoke to find Draco changing the dressing on my side.

"Mischief woke up for about ten minutes to drink some water and grab a snack before she practically fell asleep standing up. It'll still be a little while before she can tend to you."

"It's fine, Draco. I appreciate her very much." I pursed my lips. "How much did she tell you about what happened?"

"Enough to know it was quick thinking on both of your parts which got us out of that mess alive." He smiled at me. "So, you shocked me awake, huh?"

"You had stopped breathing. I actually defibrillated you."

His smile faltered. "Seriously? Mischief did not tell me that part."

"She probably didn't want to admit how scared she was."

"Cat pride," he said, shaking his head. "That must have been awful."

"Terrifying. But I didn't really have time to dwell on it with the demon advancing and all. How's Kaden?"

"Fine. He had a concussion and a dislocated shoulder, but she took care of him in the car."

"She's really amazing. Did you teach her all that?"

"You know my knowledge of the healing arts is limited, at best. She learned it all on her own. She had me bring her book after book on the subject." He brushed my hair from my eyes. "We all have our ways of dealing with heartbreak."

Anger pricked me. "She learned the healing arts, and you had your mirror."

"Angelica..."

"No. I don't want to hear you apologize again. I've had it up to here with you saying I'm sorry."

He ran his hand through his hair and stood. "I'll see if Mischief is awake yet."

"You do that."

Once he'd left, all the anger drained from my body, and I felt tired again.

Angelica?

"Hi, sweetheart! How are you feeling?"

Like that mouse I played with last week.

"I'm so proud of you. You're a real hero, little one."

Thank you. We made a good team.

She padded her way over to my bed, Draco in tow.

Draco? I feel strong enough to help Angelica. Please lift me up.

He obeyed without a word, placing her next to me on the bed.

"Are you sure, Mischief? We can wait a little longer if you want to rest a bit more.

I'm fine. Lay back, please.

With Draco's help, I complied. She bumped her head against my shoulder in greeting before snuggling up to my wounded side.

There was a slight throb of discomfort as she pressed her paws against the wound until I felt her magic working. Soon, I sat up on my own, the fight with the demon now a nasty memory and a new scar.

"Thank you, Mischief." I petted her head. "Hopefully, we won't need your talents for a little while, huh?"

Hopefully.

Draco stood as Mischief went to work healing my face.

"Well, I think we should stay here one more night and try again in the morning. Demons typically don't like to be out around dawn, so that's when we'll plan to head out."

"Sounds like a plan. Did you tell Kaden?"

"He's not speaking to me again."

"What did you do this time?"

"Nevermind, Angelica. Just rest, okay?"

"I feel fine, thanks to Mischief." I gave her another scratch behind the ears. "What's going on?"

"I'd rather not talk about it. Just drop it, please."

"Fine," I said, stung, "I'll leave you to it, then. Don't let the door hit your ass on the way out."

He turned on his heel and left, shutting the door softly behind him.

"What was all that about, Mischief?"

I was asleep for most of it. I only woke up when Kaden was yelling about someone getting hurt. Or my friend hurting someone. I'm not sure.

"Oh." My heart fluttered

The next morning, at dawn, we loaded our gear into the new SUV Draco had rented. Despite the tension between him and Kaden, at least they were speaking again.

"What's stopping them from following us?" Kaden asked.

"What do you mean?"

"Well, first there was the attack on Pepper in the hotel room, and then another the next day. If they know our location now, what's to stop them from following us?"

"I cast a heavy-duty cloaking spell on this car. I even added some rune work on the undercarriage."

"You managed that on your own?"

"You weren't speaking to me. But I handled it. If you want to reinforce it, go ahead. Can't be too safe."

We hit the road, packed and ready before the sun could peek over the horizon. We were quiet in the car. Kaden's eyes swept the scenery back and forth while Draco had a white-knuckled death grip on the steering wheel.. It wasn't until we crossed into Oregon that the tension in the car began to ease.

Draco's cloaking spell seemed effective; for the next couple of days, everything was uneventful. We cruised along

Oregon's stunning coastline, passed Mount Saint Helens, and entered Washington without incident. Soon, Mount Rainier was behind us, and we crossed into Canada without any trouble with the customs officer. Draco entered coordinates into the GPS, and we followed the directions provided by the robotic voice.

As we drove, lush green trees and sparkling lakes flanked us as we drove. Twice, Draco had to stop to allow a moose to cross. If I concentrated, it wasn't hard to imagine being the only people on earth.

After several miles, the robotic voice directed us onto an unpaved road, instantly making us grateful for the Suburban's four-wheel-drive capability. Sometimes so steep, the back of the SUV's body scraped the ground and all we could see was sky. Other times, sharp descents forced Draco to put the drivetrain in 4Lo to avoid tumbling down the hill.

After enduring an hour of the most harrowing drive of my life, the road finally began to level out. I relaxed my grip on the leather armrest and unclenched my teeth. We all took a deep breath and let out a collective sigh of relief.

"Well, that was awful," Kaden quipped. "I may have to change my pants once we get there."

"I second that," I chimed in.

"If the car has become sentient and is trying to kill us, at least her plot was foiled." He turned to face Draco. "The machines will rise another day."

"Nerd." Draco ran a hand over his face. "I think I need to pull over for a bit. That wore me out, and I need a moment to get my nerves under control."

"Whatever you need, man." Kaden clapped him on the shoulder. "That was a stellar piece of driving."

"In fact, I know we're getting close, but why don't we stop for the night?" I asked, stifling a yawn.

"I don't know if that's such a great idea, Pepper."

"Why not?"

"Because the car is warded with a cloaking spell, not us," Kaden responded.

"So we stay in the car. The seats fold down, don't they? Wait, how did you expect to take any potty breaks?"

Kaden held up a bottle.

"Ew! And what about me?"

He shrugged.

“Thanks a lot, guys.”

Draco started laughing. “He’s messing with you, Gelli. The cloak extends in a fifty-foot radius around the car. You’re safe to go tinkle as long as you stick close.”

I smacked Kaden on the arm, who was doubled over with laughter now.

“You should have seen your face!”

“Oh, shut up. I should be doubled over with laughter every day having to see your—wait, did you hear that?”

We all sat up straight.

“There it is again!” I whispered.

There was definitely the sound of something crashing through the woods.

“Is it a bear?”

“Stay in the car.” Draco opened his door and hopped down.

“Draco, be careful!” I hissed at him.

He nodded and shook out his hands.

“Kaden, stay with her.”

“Happy to.”

A tense silence filled the car as we watched Draco disappear into the trees. Occasionally, we'd hear the crashing sound, but saw nothing to explain it. No wildlife came anywhere near us. In fact, it seemed as if all the animals had vacated the area, leaving it eerily quiet. After what felt like an eternity, Draco re-emerged from the woods, bedraggled with a smudge of dirt on his face, but otherwise unharmed.

"We're definitely close," Draco said as he settled back into the car. "I found out what that sound was."

"What was it?" I leaned forward eagerly through the partition.

"A golem."

A shock of excitement went through me.

“Really? I've never seen one."

Draco opened the door and climbed in, starting the engine. "I think we'll skip camping tonight. If I'm not mistaken, we're within ten miles or so of Isaiah's cabin."

He shifted the car into drive, and we were on our way again. Peering through the window, I thought I saw pairs of eyes watching us, but I couldn't tell for sure. The absence of all normal forest sounds made the occasional crashing and groaning of trees all the more unsettling.

About ten minutes into our drive, the snow started. Small flurries dusted the landscape and danced over the windows. The flakes grew bigger and heavier, causing the boughs of the trees to droop. Draco had to switch to four-wheel drive as the snowflakes froze and spiderwebbed across the glass. Snow banks on each side of us grew until the forest was hidden from view.

Eventually, the Suburban slid and came to a stop, embedded in snow. The entire right side of the SUV was covered, blocking us in on one side.

"Isaiah has been known to control weather as well," Draco sighed, shutting off the engine. "We're going to have to hike the rest of the way."

I shook my head. "In what clothes, Draco? We're hardly in winter wear."

"Check in the back. I had a suspicion we may encounter something like this and brought along some extra gear."

I climbed to the third row and hoisted an army duffel to the front. Draco reached over to open it and started pulling out North Face and Gortek clothing, along with snow boots.

"You weren't kidding." Kaden ran his hand over one of the puffy down jackets. Draco nodded.

Now, as we prepared to trudge through feet of snow, the cozy warmth of the car became difficult to leave. But with my life at stake, failing to reach Isaiah was not an option. I started layering on the clothing Draco had packed for me.

Once I had my boots on, I hopped out of the SUV, immediately sinking up to my knees. A snowball hit me on the shoulder. I turned, and Kaden shrugged.

"May as well."

I narrowed my eyes at him and brushed the snow off my shoulder in slow, deliberate strokes. Bending down, I quickly packed a snowball. Mine hit him square in the chest while another glanced off my hip. I dodged one aimed at my head from the left and whirled around to glare at Draco.

"Really? Ganging up on a girl? What grade are you people in?" I harrumphed. "Fine." I cast a quick succession of runes and created a thick shield of fire around me, giving them both a smug smile. "Get through that, you asshats."

With a sexy twist of his lips, Draco turned and nailed Kaden with the snowball originally crafted for my benefit. I rolled my eyes.

"Boys," I muttered under my breath.

Beneath my feet, the snow slowly melted to reveal solid ground and green grass. I watched with a smile as small flowers sprang back up, stretching their leaves as if yawning.

"Now, what is this…?" I kneeled down.

Between two piles of leaf litter was a small, glowing symbol. Ignoring the laughter around me, I brushed the debris aside to find intricate rune work glowing in the dirt.

"Hey, guys?"

"You can't stop this barrage, punk!" Kaden whipped up a spell to send multiple snow missiles at Draco. He dodged and threw fire spells, exploding them in midair and managing to get hit by only half of them.

"Guys?" I kneeled to get a better look before turning their way.

"You think you can take me down?? My name is Winter!"

"Guys!" They both stopped and looked at me. Draco trudged over and stood above me, cheeks flushed.

“What’s up?” he asked, catching his breath.

"Take a look at this." I dropped my fire shield, and he crouched down beside me.

"Wow. That’s some high-level stuff."

"My thinking exactly," I replied as Kaden joined us.

"What are we looking at? Woah…" He reached down, his fingers hovering inches above the runes. "It’s masterful."

Draco rocked back on his heels and let out a breath. “I think we're very close to Isaiah.”

Suddenly, something pinned my arms and lifted me from behind. I let out a short scream, kicking my feet in the air. Kaden and Draco whirled around.

“Behind you!” I shouted, too late.

Each was grabbed around the middle by seven-foot-tall humanoids. They appeared made of trees and clay but moved with fluidity and had glittering eyes. A freezing wind kicked up, cutting through my heavy jacket and stinging my face. The boys struggled uselessly.

“Enough,” said an unseen voice mildly.

I was dropped unceremoniously to the dirt. The boys followed shortly after. A teenaged boy dressed in a long robe stepped from behind a tree.

“So you’ve found me and my creatures.”

“Isaiah?” Draco asked, brows furrowed.

“Yes.”

Isaiah turned and began walking away. The creatures followed. After a few yards, he stopped and turned back around.

“Well, are you coming or not? I imagine you’ve traveled a good distance to meet me.”

His voice broke through our shock, and we started following him. Isaiah led us to a path which materialized as he approached. He brushed aside some branches, holding them for me as I passed.

"Keep going. I’ll catch up in a moment.”

He turned to face his creatures, lifting his arms. Speaking ancient words of power in low tones, the golems turned around, likely returning to their previous post at the entrance. Passing us by, he ignored our astonished expressions.

We continued in silence, observing as Isaiah brought us through half a dozen wards. He would stop, tilting his head as if listening for something, then trace sigils in the air to disarm whatever protections he had set up, allowing us passage. He would repeat the process in reverse once we'd gone through.

The weather shifted to rain, melting the snow into puddles. The green of the trees regained its vibrancy. The woods echoed with bird calls and chittering squirrels once more. As we walked, the air grew warmer around us, and the rain stopped, prompting us to shed our jackets. Flowers bloomed before our eyes, and bees wandered lazily from blossom to blossom. I took a deep breath and it tasted like Spring.

"Isaiah… This is incredible."

"You are the first people to see this in a long, long time, Angelica."

"You know who I am."

"Of course."

"So you know why we're here."

He paused, cradling a rosebud which opened in his hand.

"Yes," he said quietly, "and I can't help you."

My heart sank into my stomach.

"What?" It came out as a whisper.

He turned to face me. "I can't help you. I will give you safe passage now and provide you a place to rest and plan. But I cannot help you."

"What do you mean you can't help?" Draco advanced, fists clenched.

Isaiah matched him step for step until they were almost touching. He looked young and small with Draco towering over him.

"You wish to fight me, mage?" He lifted an eyebrow. "That would be unwise."

Draco unclenched his fists and took a deep breath. "No. But you understand—"

"I understand much, mage. But this is not my war. I do not belong on one side or the other."

"That's bullshit." Draco took a step back and understanding dawned on his face. "You're afraid."

"Nonsense. I fear nothing."

"You're afraid of Damien."

Isaiah took a step forward and the grass at Draco's feet began to smoke.

"Draco," I said softly.

"As I said, I fear nothing, mage!"

A shrub next to Kaden burst into flame, causing him to jump.

"Draco!"

Draco was squaring off with an ancient. It was not a fight he could possibly win.

"Stop it, both of you!" I shouted.

Isaiah stepped back, and the shrub returned to its bushy, green state. Small flowers dotted the grass around Draco's feet. I let out the breath I'd been holding, and the tension in Kaden's shoulders eased a bit.

"Now, if you'll follow me, I'll have a meal laid out for you."

He waved his hand, and a door appeared in front of us. Isaiah turned the latch and walked in as if it were just another sunny afternoon in the forest.

I turned to Draco.

"You have to back off. He looks like a kid, but this is one of the most powerful beings in our world. Don't get yourself killed over pride."

He started to say something, but I cut him off.

"Don't get yourself killed for me." I was surprised by the tears in my eyes. "Please."

"Gelli, we need him desperately. We will lose without his help."

"I understand that. Believe me, I do. But this swing and measure standoff will only get you killed."

I turned on my heel and walked through the door where Isaiah had disappeared.

Kaden placed a hand on Draco's shoulder.

"C'mon, man. Let's get some food and plan. Just like the kid suggested."

Draco nodded and entered the house without another word.

Stepping over the threshold of Isaiah's house was stepping into another world. A tingle of magic washed over me as I passed through the warded doorway, and the forest behind us disappeared. Gone was the humble cottage in the woods. A grand foyer opened into a large room rich with mahogany and lush antique furniture. Books lined every wall, and tomes stood in towering stacks on various tables. I craned my neck to try and see where the bookcases ended, but they stretched up beyond the reach of the lamps lighting the room.

"You may hang your jackets in the closet to your left. Please take a few moments to freshen up. You'll recognize your rooms when you see them."

We did as he bade us and hung up our coats before shuffling down a hallway. I puzzled over his cryptic comment, wondering how we would recognize rooms we'd never seen before.

Lunch And Learn

"What do you mean you lost her?!" Damien screamed into the phone.

"Sir, we had them at the hotel—"

"I know you had them at the hotel," he said through gritted teeth, gripping the phone until it creaked in his hand. "And you let her get away! Now you're telling me you don't even know where she is!"

"She had two powerful mages with her—"

"Excuses!!"

Damien growled in frustration and hurled the receiver against the wall, shattering it into plastic shrapnel. Books crashed to the floor, and a paperweight embedded itself in the wall as he swept his arm violently across the table. He strode to the front door and slammed it behind him, stomping a path to the sports car in the garage. The cheerful chirp of the alarm deepened his scowl.

Briefly closing his eyes. He engaged the clutch and fired up the engine. It roared to life, and he slammed it into gear.

He had work to do.

~

"Do you guys hear that?" I asked the boys as we made our way through the house. A low tone hummed in my ears.

"The music?" Kaden asked, tilting his head.

"Yeah, I hear singing." Draco paused his steps and placed a hand on the wall. Kaden regarded him pensively.

"No, not singing. I hear the piano."

"I hear a humming."

They turned and looked at me.

"Right then," Draco said, scrubbing his hands over his face. "Shall we?" We nodded and continued on our way.

The humming grew louder until my chest vibrated and my teeth buzzed with an oddly pleasant sensation. I stepped up to a heavy purple door adorned with a large orange "A." The humming stopped.

"I guess this is me."

Draco reached for the doorknob, which shocked his fingers. I smiled.

"And no boys allowed, apparently."

"I didn't even feel the ward. His magic is subtle even if the paint is not." He rubbed his hand.

I reached for the door, and it opened with a click. Soft, gilded light spilled out with the delicate scent of peppermint. I breathed deeply, closing my eyes.

"You know, for a recluse, Isaiah sure knows quite a bit about us."

"Meaning?" Kaden peered over my shoulder into the room.

"Our names, for one. Why we're here. My favorite color combination and my favorite scent. I'm also guessing our meals will be catered to each of us."

"It's who he is," Draco said with a shrug, "he's a collector."

"Well, that's not alarming," Kaden said with wide eyes.

Draco shook his head. "Not that kind of collector. Just imagine being one of the first immortal beings."

Kaden's eyes grew wider. "I honestly can't."

"What else is there to do after a while but read, listen and learn? He collects information."

"I suppose that does make sense." Kaden ran a hand across his mouth.

"Well, guys, I'm going to take a quick shower before we eat." I said, wanting to shelve this conversation until I'd had a chance to wash the road trip and the weird scent of golem from my skin.

I released my hair from its ponytail and shook it out. I saw a muscle twitch in Draco's jaw as he swallowed hard.

"Let's go find our rooms," he said gruffly, walking away.

Kaden shot me a sympathetic glance before turning to follow. I watched them for a moment, then squeezed my eyes shut and ran my hands roughly through my hair. If I didn't know better, I would have thought Draco was resisting an urge to run his fingers through my hair. Maybe take me in his arms, grab the locks at the back of my head in his fist before kissing me like his life depended on it...

Letting out a breath, I shoved those useless thoughts into a mental box before removing my shoes and socks. I rubbed my toes in the soft carpeting while I took in the place.

The room was decorated in an autumnal palette with subtle purple splashes. Sheer gold and copper drapes flapped in lazy ripples around French doors, open just enough to let in a cool evening breeze. The twilight of the waning day crept in and bathed the floor in lavender hues.

My feet sank into the plush, moss-colored carpeting as I walked the space, investigating the subtle decorative touches. The whisper of an occasional breeze was the only sound in the room. I stopped in front of a wall-sized tapestry depicting the Queen of Swords. The detail was so fine I could practically feel the sharpness of her blade, see the fierce wisdom in her gaze. I traced the back of my hand along her face and crown before shedding the rest of my clothing, draping it over a chair, and heading to the shower.

The waterfall shower was such bliss; I had soaped and rinsed twice before after seriously considering changing my address, I willed myself to get out. A cloud of steam billowed behind me as I opened the bathroom door. I looked up and stopped in my tracks.

Nestled in one corner and covered with decadent fabrics and overstuffed pillows was now a massive four-poster bed which had not been there before my shower. Each poster of the frame was a carved replica of the pillars of Ibrahim and Boaz, the columns which flanked the High Priestess in the Rider-Waite deck. Complementing the pillars was Hebrew and Aramaic lettering. The headboard featured a magnificent relief of The Magician.

Oof.

~

The Crone

Damien pushed the car to its limit, redlining it most of the way. He ignored the highway patrolmen who attempted to catch him, mangling their cars with a wave of his hand. He didn't have the luxury of time.

He needed to reach The Crone.

He raced South from the Midwest, disgust and fury raging in his chest. Glowering at the flat landscape of fields and crops, he gnashed his teeth at his stupidity for employing inferior, low-ranking demons to accomplish his goals. The tires squealed in protest as he screamed around corners. Soon enough, he would be in the South, one step closer to the one he sought.

It was a dangerous gamble to seek out The Crone. She had no alliances to save herself, and there was a decent chance she would refuse to lend her skills for no discernible reason. There was also the possibility she would actively and painfully reject him altogether. But she was the only one he was certain could achieve what he had in mind. He continued his drive south.

He reached the border of Louisiana as dawn crept over the horizon. He willed himself to release his anger. She would not even agree to see him if he came on too strong. Marshes flanked him, their eddying pools of water tugging at the branches of massive weeping willows. He slowed the car. Under different circumstances, he might have found it peaceful. Black eyes winked at him and disappeared into the murky water. His lips twitched into a smile for the first time in days. Predators were his kith and kin.

The moon rose as his destination drew nearer. Pulling the car to the side of an abandoned road, he shut off the engine, straightened his cuffs, and stepped out into the humid night.

Oppressive as the air felt, his steps were light as he picked his way through brambles and vines. Cicadas harmonized with nocturnal creatures, and fireflies around him blinked in and out of sight.

After walking a half-mile, the trees and vines closed in, and Damien began to feel the tremor of magic. Navigating through thorns and brambles, he parted a curtain of hanging vegetation braided with spiderwebs. Finally, he encountered The Crone's perimeter ward. This bit promised to be unpleasant. He took a deep breath and sliced a deep gash in his palm with the consecrated knife he had brought for this purpose. He pressed his hand against the ward.

A shock of pain reverberated up his arm, and instinct almost had him pulling away. He gritted his teeth and kept his skin pressed against the barrier, feeling her magic seep, abyssal and oily, under his skin. After ten minutes, nausea crept in, and the smell of burning flesh filled his nostrils until he could smell nothing else. Sweat stung his eyes. His breath came in short gasps. His vision tunneled, and the sky turned stygian.

A few minutes later, his knees buckled as the agony intensified. He was consumed with an overwhelming urge to retreat, desperately battling the voice screaming at him to sever the connection and run. He clenched his jaws so hard his molars squeaked. With a pained grunt, he focused on his motivation for seeking out The Crone and pressed his scorched flesh harder against the barrier.

Suddenly, the resistance gave way, and he fell forward, panting, onto his hands and knees. Rolling to his side in the mud, he cradled his hand to his chest and curled into himself, letting out a groan. Shuffling footsteps approached behind him. A chill ran over his skin, and the hairs on the back of his neck stood up as breath, putrid and raspy, whispered against his ear.

"Well done, child."

"Thank you," he replied politely before passing out.

~

Damien awoke shivering on a stone hearth in front of an unlit fire. He struggled to turn his head toward the slow creaking sound of a chair rocking over old floorboards. He blinked his eyes into focus. The Crone regarded him silently from a spot in the corner, eyes narrowed, a pipe dangling casually from her left hand.

"I'm certain you wouldn't risk your life coming here if you didn't have a reason, Damien."

His limbs aching from the cold, Damien flicked his eyes to the fireplace and considered igniting the stack of firewood. The temperature in the room dropped further.

"It would be unwise to attempt magic in my home, demon."

His joints screamed in protest as he willed himself to a sitting position.

"I would not presume."

He held back a gasp as pain lanced his spine.

"Lie to me again, and the end of this meeting will be abrupt and painful for you."

Damien clamped down on his outrage. He dipped his chin respectfully and dropped his gaze to the floor before him. He needed to play this right, or he could lose everything. He caught a shimmer in his peripheral vision and chanced a glance at the corner where The Crone sat.

Her visage had changed from a haggard old creature to a devastatingly beautiful young woman. A pristine white sundress hugged her curves, ending just above her knee. She crossed her legs, flashing him a glimpse of a perfectly sculpted thigh.

Damien schooled his expression. She had a vision which matched his ideal in every way. A deliberate move and a decent one on her part. Her long chestnut hair cascaded in gentle, shining locks and tumbled over the creamy skin of her bare shoulders. Impossibly green eyes now speared him, sparkling with mild amusement.

She languidly dropped her hand over the arm of the chair, and a panther slid to her side with a low growl. Damien's heart flipped in his chest. The Crone's familiar Siren sat back on her haunches with deadly grace, muscles rippling. The Crone rested her hand on the cat's head, stroking behind her ears lovingly.

Damien swallowed, trying to ease the dryness in his mouth. Old as The Crone, Siren was as dangerous as the witch herself. Possibly more so. No living creature knew the full power of the cat save her master. Speculation was they were two halves of one being or, possibly, Siren was The Crone's avatar. Either way, Damien was in no hurry to test any theories while sitting at her feet on a cold stone floor like a supplicant.

He tried once again to rise and found his body would not cooperate. She raised an eyebrow, and a second rocker materialized beside him. With a subtle nod toward the chair, she indicated he sit. Mustering the remaining shreds of his dignity, Damien pulled himself up, shoving down the irritation which threatened to rise at her treatment of him. He eased into the chair with a mental groan. She chuckled softly.

"Do not criticize yourself too much, demon. You should be proud. Pain and discomfort are woven into the very fabric of my home for the sake of any guest. Few have made it this long in my house without succumbing to madness." A smirk flitted across her lips. "Or death."

"Fortunately for me, I've been unconscious for most of it."

She flashed white, even teeth.

"Fortune has nothing to do with it." She stopped petting Siren and folded her hands delicately in her lap. The cat watched him with eyes as impossibly green as her master's. "Very well then," she said, leaning back in her chair.

"I'm listening."

~

Dinner was quiet and thick with tension. Draco pushed at the egg foo yung on his plate. His lips pressed into a thin line. After a half-hearted attempt at my breakfast food, I mentally shrugged and pushed back from the table. Kaden and Draco both looked up at me.

"Leave it," Isaiah said as I reached to collect my plate. He cut into his steak casually, his eyes fixed on the meal in front of him. I withdrew my hand before standing and replacing the chair. I turned to leave when Draco cleared his throat.

"I just have one question for you, old man."

My shoulders slumped. *Here we go*, I thought with a mental sigh.

Isaiah laid down his cutlery and angled his body to face Draco, leveling him with a cool stare. Draco matched his posture. Kaden wiped his mouth with his napkin and carefully pushed his chimichanga away.

"Which is?"

"You say you are unafraid."

"That is not a question."

Draco narrowed his eyes.

"If it is not fear holding you back from helping us, why do you refuse to do so?"

Isaiah turned back to his meal, taking up his knife and fork again. "This is not my fight," he replied, popping a small piece of steak into his mouth.

Draco slammed his hand down on the table so hard the dishes rattled. Isaiah stopped chewing.

“That’s not good enough!

Isaiah finished chewing, washing it down with a swallow of claret. Suddenly, he stood and, moving faster than I could track, fisted Draco’s shirt and lifted him from his chair. He spun, took two swift steps forward, and pinned him to the wall, feet dangling. My heart leaped to my throat. Rage warred with uncertainty in Draco’s eyes as he gazed silently down at Isaiah.

“Do not presume to know me, mortal,” Isaiah growled in a voice too low to be entirely human.

He lifted Draco higher, pulling him away from the wall and slamming him back again. Kaden was already standing, energy arcing across his fingers. Draco glanced at him with a subtle headshake. Kaden fisted his hands at his sides and took a step back.

“Put him down,” I whispered.

“You are afraid,” Draco said in an even voice. “Throwing a tantrum will not change that.”

Isaiah gritted his teeth, an animal sound reverberating from his chest.

“Put him down.” My skin heated.

"Damien will raze the earth looking for her." He nodded in my direction, his eyes trained on Isaiah's. "You think declaring neutrality will save you?"

Isaiah roared and gripped Draco's throat.

"Put him down," I repeated, fingers twitching.

"I have seen civilizations rise and fall more times than years you have been alive, child. I have watched the world burn about my feet and remake itself in countless ways. I've witnessed the oceans flood the earth, dry to a desert, then fill once more. This planet, this existence, was birthed before my eyes, and you presume to speak to me as if you know my mind??"

Draco's lips opened silently, his face turning an alarming shade of crimson.

"Put. Him. Down!" Energy burned in my core and I felt my hair lift from my shoulders. "NOW!"

A groaning rumble shook the house and the lights in the kitchen popped above us. The dishes jumped and fell to the floor with a crash. Glass shattered and wine mixed with beer and diet Mountain Dew to stain a mosaic on the tablecloth. Isaiah's eyes moved casually around the room before he dropped Draco to the floor where he coughed and gagged, heaving great gulps of air.

Isaiah turned my way, eyebrows raised. His expression morphed from rage to keen interest so immediately it was disorienting. He stroked his jaw and scrutinized me for so long I had to fight the urge to fidget. Other than the sounds of Draco's agonized attempts to breathe normally, silence reigned.

Isaiah strode past me without glancing at Draco's prostrate form or Kaden's open-mouthed stare. He hesitated by the doorway and turned to regard me again, his expression now inscrutable.

"We have things to discuss." He turned back around and called back, "We shall speak in the morning."

And with that, he was gone.

The boys and I stared at each other, blinking.

"Well," Kaden said after a moment, "He's a peach, isn't he?" I ran over to Draco and kneeled beside him, where he lay on the floor.

"I mean, maybe a touch dramatic…"

I shot Kaden an exasperated look. He shrugged.

I placed my hand on Draco's shoulder, cupping his face with the other. His shuddering breath brushed across my palm and warmed my wrist. I dipped my head until my eyes were level with his.

"Are you okay?" I asked softly.

A crease formed between his eyes, and he cleared his throat. He swallowed visibly and nodded.

I reeled back and slapped him. His mouth made an 'O' of surprise.

"What the hell were you thinking??" I went to slap him again, but he caught my wrist, so I pounded on his chest with my free hand instead. "He could have killed you!" Frustration and anger boiled over until my eyes blurred with tears and my voice cracked. "Then where would we be?? Where would I be?? What would I do without you??"

The moment the words left my mouth, I froze. Draco dropped my wrist, and I did the only thing I could think of.

I ran.

~

The Crone maintained her maiden form as she moved gracefully through her house, opening cupboards and gathering supplies. Watching the soft, white dress play over her perfect skin, Damien felt his arousal stir. She paused what she was doing to toss him a sultry smile over her shoulder. He shifted in his chair. She bit her bottom lip and ran a hand through her hair.

"This form tempts you."

A new game, then. Damien bit his tongue until he tasted blood, swallowing his reply. He cleared his throat.

"I appreciate beautiful sights, and you are most certainly that."

"Hm."

Between one thought and the next, Damien was suddenly overcome by an unbidden desire for her so acute it weakened his knees and hardened his cock until it strained his pants, throbbing painfully. With a small smile, she arched her back in a stretch, and the white fabric strained over her perfect breasts. She slowly bent over with a deep sigh to tend the fire, her skirt riding up to reveal more of her delicate skin, her cleavage spilling sumptuously over the hem of her sundress.

His throat clicked as he tried to swallow over the sudden dryness in his mouth. He stood and took two stuttering steps toward her before forcing himself to freeze. She straightened, turned his way, and walked slowly to where he stood, her heavily-lidded eyes peering up at him. Her small, pink tongue darted out and wetted her lips.

The room suddenly felt too small. She rose to her tiptoes and leaned forward until the fabric of her dress tickled his bare chest.

Wait. Where the hell is my shirt? Reason clutched at his lust-addled brain.

"What is it you want, Damien?" She whispered against his mouth. Her breath tasted of honeysuckle.

Answers tumbled around his head, the foremost coming straight from the aching erection straining against his pants.

"I want…"

"Yes?" She purred.

Like a thunderclap, Damien seized control of his mind again. He growled in frustration before clothing himself in a supremely expensive suit woven from threads of energy directly from his father's kingdom. Siren responded with a growl of her own from her spot beside the fire.

"Have it your way, demon," she pouted.

"Enough games, Calliope," he ground out through clenched teeth.

She narrowed her eyes at his use of her real name. Siren rose with liquid grace and circled him. Keeping the cat in his peripheral vision, he seated himself back in the chair. Now came her Game of Wills. He bit back a scowl, his frustration beating against the walls of his control. His patience was wearing dangerously thin.

"Careful, demon," she said, her voice taking on a razor's edge.

Then she pursed her lips into a dainty pout and shrugged. "Fine. You may tell me why you are here."

Damien held back a sigh.

Finally.

~

I didn't slow until I reached my room and shut the door firmly behind me. I was mortified. How could I say that? Why? My words echoed in my head on an endless, torturous loop. I felt so stupid and weak. This man shattered my heart, and now I was practically throwing myself at him.

There was a soft tapping at my door.

"Pepper?"

"Go away, Kaden." I couldn't face him. He must've thought me such a fool.

"C'mon, Pepper. Let me in."

"Hard pass."

There was a brief silence.

"I have bacon…"

I considered resisting but knew it was a lost cause. I walked over to the door and opened it. He stood on the other side with a napkin-covered plate in one hand and a cup of coffee in the other. He grinned and strode past me into my room, placing the dishes on a writing desk beside the French windows. I reached for a piece of bacon, but he moved the plate out of my reach.

"Hey!"

“No bacon until you talk.”

I crossed my arms and narrowed my eyes at him. “No fair.”

He shrugged.

I glowered at him a little longer before breathing out a resigned sigh. “Kaden, I’m so mortified. I can’t believe I said that.”

“You know I would never judge you.”

“I know. Not to downplay how important your opinion of me is, but your feelings on what just happened are not at the forefront of my mind right now.”

“You wound me.”

I rolled my eyes and tried to sneak the bacon faster than Kaden could react. He snatched the plate and lifted it above his head.

“What are you, twelve?”

He smirked at me. “Keep talking, and maybe you’ll get to enjoy a strip of bacony goodness before they get cold.”

“You’re mean.”

He shrugged again.

"Okay." I sat down on the edge of the mattress and hung my head. He sat beside me and placed the plate between us.

"He hurt me so deeply, K."

"I know. I was there, remember?"

I nodded.

"The thing is…"

Kaden watched me silently with his big, brown eyes, waiting for me to continue. The words stuck in my throat. I shook my head at myself. My stupid, stupid self.

"How can I still love him, Kaden?" I murmured quietly.

He responded with a soft, sad smile and draped an arm over my shoulders.

"Pepper," he began gently, "I don't pretend to know anything about being in love, having never experienced it firsthand. All I know is my parents who left, your parents who are psychotic, and you and Draco." He gave my shoulders a light squeeze. "The thing is," he continued, "I don't think you and Draco - the two of you together - is such a poor paradigm to aspire toward."

My mouth fell open.

"Are you joking? He cheated on me, Kaden! Cheated!"

"Yes, Pepper, I remember. I'm not going to tell you what to do. I have no idea whether something like that is even forgivable. Only you can decide that." He took a deep breath and angled his body to look at me. "But, for your sake, I want you to ask yourself this - and be honest with yourself. Are you truly still hurt and angry?"

"How can you even ask that, K?"

"Let me rephrase. How much of it is pride and fear?

"I'm not afraid of him."

"I didn't say you were afraid of him. I know you are not. No, I believe you fear being vulnerable again."

I felt the color drain from my face, and my fingers went numb.

"You're wrong."

"Okay."

I glared at him.

"Like I said, I know nothing about being in love. If you say you are not afraid of opening yourself to the risk of having your heart shattered for a second time, I believe you."

He withdrew his arm and squeezed my hand before standing. He made his way to the door and paused. Without turning around, he said,

"You deserve to be happy, Pepper."

Then he left, closing the door behind him. I could hear the pounding of my heart in my ears, deafening in the quiet of the empty room.

I reached for a piece of cold bacon.

~

The Crone flashed Damien a pleased smile once he had divulged what he wanted from her. It ticked all her criteria to be deemed worthy of her talents—a challenge was enough so she could stretch her abilities in addition to being marvelously creative, allowed creativity, and above all, promised exquisite cruelty. It was this last aspect which had earned him the smile.

She even gave him the rare option to stay while she worked, an offer she had never extended to anyone. He tried to disguise his intrigue at the prospect, but she picked up on it nonetheless, and her ego preened and crowed.

She found him fascinating, attractive even. Thoughts of adding a carnal clause to their agreement crossed her mind fleetingly, but she dismissed the idea. She didn't need a contract. Soon enough, he would be begging for release from the torment she had in store for him.

But that was for later. Now, it was time to get to work.

She began gathering the necessary materials earnestly, not just for show as she had been. There was much she needed to construct. Constructing such an intricate manifestation required careful preparation.

"I don't suppose you've brought anything of hers, have you?"

"Regrettably, no."

"Hm. I'll still be able to do it. I'll just need to get a bit resourceful."

She didn't actually require anything of the girl, but she wanted to impress him even more with her ability than he already was. Now, she'd set her sights on him as a future lover, laying the groundwork to establish herself as the goddess he would soon consider her. Otherwise, he may get the idea they were equals, and training him properly would be tiresome.

Pushing those thoughts aside for now, she focused on the task at hand. She shot him a playful wink over her shoulder and got to work. Damien resisted the urge to roll his eyes behind her back and settled with crossed legs, observing her from his corner of the room.

She closed her eyes and ran her hands along the table, feeling the belladonna and brambles under her fingertips. She crushed them between her palms. Next, she added sap from poppies grown in her garden. She placed the ingredients in a large mortar and pulverized them with her pestle. Pricking her finger, she squeezed three drops of blood into the mixture. Then came the paste made from spider's eggs and nightshade.

Blending the ingredients into a thick paste, she retrieved her obsidian athame and dipped it into the concoction. Once the knife was coated, she used it like a pen, drawing a precise cube on a fox hide with her handmade ink.

"Now, I must meditate. This is the most delicate part of the spell, so please remain still," she said, sitting in a cross-legged position on the floor.

Closing her eyes, she reached for the ley line running beneath her house. Threads of power wrapped around her fingers as she began weaving the construct. She swayed slightly as she weaved, her eyes darting behind her eyelids. Her hair floated around her as if she were underwater. Her body lifted from the ground. Damien watched with intense interest. He grudgingly admitted to himself he found her alluring. Being permitted to witness her craft was captivating.

A breeze stirred within the house and the temperature dropped. The candles she had lit winked out. Next went the room's lights, plunging the room into darkness. Her skin began to glow with such intensity which Damien had to squint. A low, primal sound rumbled from her chest, a cross between a hum and a growl. The wind picked up, pushing against Damien in his chair.

Her body stiffened suddenly, eyes snapping open. Her irises had vanished, leaving only milky-white orbs rimmed in black. Her head jerked back, and her mouth fell open silently. Her body jerked awkwardly until her back arched to an unnatural angle. A sickly yellow corona engulfed her, and her hands began to char, sending plumes of and emitting thick, black smoke into the air.

Damien remained silent, suppressing a grimace as the air filled with the scent of his father's kingdom and something dark and unsettling he couldn't place. He narrowed his watering eyes and forced himself to remain still. Finally, a loud crack shook the room, followed by a boom which rattled the windows. His ears rang as she gently floated back to the floor once again.

She lowered her arms, and the lights flickered back on. The fireplace roared to life. Turning to face him, flames from the inferno danced and flickered across her features.

"It is done."

Damien flashed her his most dazzling smile.

"Brava, beautiful one."

~

The second time there was a knock on my door, a flash of panic surged through me. I considered staying quiet and feigning sleep, but masochism drove me to let Draco in. I swung the door open and turned on my heel, assuming he would follow.

Without looking back, I made my way to the bed and flopped onto my back, screwing my eyes shut to avoid looking at him. The bed dipped as he sat beside me. The closeness felt too intimate, so I shot up and scooted away until my back hit the headboard. I hugged my knees to my chest and rested my forehead on my arms.

"Angelica?" The bed shifted again, and I felt his eyes on me. I shook my head, rocking it back and forth against my forearm. He sighed.

"I'm just going to talk, okay?"

I shrugged and made a noncommittal sound.

He took a deep breath, and I mustered the courage to lift my head. If possible, he was even sexier than ever. His hair was tousled, and his hazel eyes were luminous. They always turned a gorgeous color when he was upset—moss green and gold. I felt my cheeks flush. He cleared his throat.

"First, let me say I'm sorry for scaring you. It was selfish. I wasn't thinking, and I let my emotions get the better of me."

He paused, giving me the chance to speak. I stayed silent.

"The fact is, I'm terrified."

I inhaled sharply. His gaze dropped to my mouth before he quickly shut his eyes. When he opened them, he trained them back on mine.

"It's just…" He scrubbed his hand over his mouth and turned to face me fully. "I won't insult you by apologizing again for what I did to you, to us, twenty-seven years ago. I had no excuse then, and I have none now."

"Draco…"

"No, let me finish. If I don't get this out, I may never have the guts to do it again."

I pressed my lips together and nodded.

"I want you to know I haven't been with another woman or even dated since you left."

My jaw dropped.

"Not one?"

"Not one."

"Draco... That's insane. Why on earth would you punish yourself like that? We were over. You could've had any woman you wanted. As many as you wanted." Even after almost thirty years, the thought of him with another woman made my stomach twist with jealousy.

"Not any woman," he said softly, studying the bedspread.

"What are you saying, Draco?"

"There is only one woman I want. Only one woman I've wanted since the day you left me."

He raised his head to search my face imploringly. The naked emotion and sincerity in his gaze almost made my anger dissipate. Almost.

"So, what? I leave and suddenly become the only one you want. Why? Because I'm unavailable and therefore a challenge? Just wanting me because it's human nature to covet what we cannot have? That's such bullshit, Draco. What you did destroyed me! It almost killed me!"

He nodded. "I know."

"Oh, that's right. You invoked a voyeur spell to watch me like a stalker."

He shook his head. "No, you've got it wrong."

"Do I? Then enlighten me."

"I didn't invoke the voyeur spell to stalk you like some obsessed, deluded fool suffering from unrequited love. I did it because I realized I would never see you again after what I did. To go the rest of my long life without seeing you. I knew I would go mad with grief. Jesus, Gelli. Within weeks after we broke up, I did the only thing which made sense to a desperate man wracked with guilt and heartbreak, staring at decades, centuries without the only person I couldn't live without. I decided I couldn't face another day, knowing I had ruined the only thing in my life that gave it meaning. That I had hurt the one person who meant more to me than my own life."

I felt the blood drain from my face. "What did you do, Draco?" My voice came out just above a whisper. Fear clutched my heart.

"I tried to kill myself, Gelli."

An anguished cry escaped me. I felt as if I'd been punched in the gut. Then he continued, and I almost wished he hadn't.

"Actually, I succeeded."

"No…" Tears welled in my eyes, but my throat was so constricted they wouldn't fall.

"The voyeur spell didn't occur to me until...After."

I shoved my fists against my eyes until I saw stars. I managed one hoarse word. "How?" I looked at him again and saw the pain in his eyes.

"What do you mean? As in the method?"

"No, asshole! I mean, how would you be here if you succeeded?"

"Mischief."

Realization dawned. "That's when she started learning the healing arts, isn't it?"

"Yes."

"You selfish prick. How could you leave her like that? She's your familiar, Draco! That would've been a death sentence for her as well!"

"I know," he said in a strangled voice.

He sounded so broken. Sinking from the bed to the floor, he drew his legs up and propped his arms on his knees. He dropped his head.

Then he did something I had never seen before, causing me to freeze.

He began to weep.

I unfolded my legs, got off the bed, and stepped around until I stood just above him. His shoulders shook, and the pain of his cries melted the last bit of ice in my heart. Kaden's words about pride and fear returned to me, and it all clicked.

I reached down and ran my hands over his shoulders and neck. Snaking one hand behind his head, I buried my fingers in his hair. He rolled up until he was on his knees in front of me. He gripped my shirt and pulled himself into me, sobbing against my stomach.

"Gelli, I'm so sorry," he choked out.

"Shh, Draco."

I let him weep against me, stroking his hair gently. My eyes fluttered closed, the feel of his body fitted against mine intoxicating. He tugged on my shirt until I sank to my knees, facing him. He wrapped his arms tightly around me, one hand cradling my head, the other splayed across the small of my back. I continued running my fingers through his hair until his sobs began to quiet.

He took a deep, shuddering breath and raised his head. I automatically cupped the side of his face with my hand and wiped the tears from his cheek with my thumb.

I inhaled his achingly familiar scent. Like fresh air and rain-soaked earth. Something undefinable and unequivocally masculine. He smelled like home.

He rocked back onto his heels. The move forced me to let go of his hair. Reluctant to stop touching him, I ran my fingers down his neck to rest my hands on his strong shoulders. He looked at me with such desire, a gasp escaped me and a pulse of need shot through my core. His gaze dropped to my lips, and his pupils dilated. My heart pounded in my chest.

He leaned in at a torturously slow pace, giving me every opportunity to stop him. I felt his breath inches from my lips when he paused.

"Gelli?"

"Yes?"

"May I please kiss you?"

My heart leaped into my throat, beating a frantic staccato in my ears.

"Yes," I whispered.

After so long, I expected something feverish and demanding, but he closed the distance with a light brush of his lips across mine. Something between a sigh and a moan escaped me, and he smiled.

He kissed me again, this time more firmly. He ran his tongue along the seam of my mouth, seeking permission. I let him in, and his tongue stroked mine sweetly. Reverently.

So much time was lost. Years wasted. The pain I'd lived with for so long, worn like an ill-fitting suit, began to recede. I was once again becoming whole.

Desire bloomed hot between us. He ran his hands up my sides, and goosebumps rose along my flesh. His fingers brushed the sides of my breasts on their way to tangle in my hair.

When we broke the kiss, we were both breathing heavily. Suddenly, he grabbed me, and I fell into his lap with a squeak. He grinned and buried his face into the crook of my neck, holding me tight. He said something, but it was muffled by my shoulder.

"What was that?"

He shifted until his forehead rested against me.

"I've missed you so much."

I squeezed him.

"I've missed you too, Draco Winter."

He tucked a lock of hair behind my ear before tracing a slow, soft trail down the nape of my neck to my shoulder, trailing fire in its wake. Need overcame me in crashing waves. He asked me something, but sitting in his lap with his hands on me was so distracting I didn't hear the question. He ran his fingers down my arms, barely touching my skin, and my lips parted with a sigh.

"Hmm. That feels really nice."

He placed a gentle kiss on my collarbone, and I tipped my head to give him better access. He trailed kisses up to my jaw, and I felt him smile against my skin. He tugged me by the hip with one hand while the other continued its exquisite torture. I could feel his length hardening beneath me, and a pulse of desire shot through my core.

"Draco…"

"So, is that a yes?" He dropped his hand and held my hips possessively, gazing up at me.

I looked down at him, basking in the heat of his need. The desire in his eyes stoked the fire between us. I licked my lips.

"Mmm-hmm. Definitely, yes."

"Okay." He popped open the first button of my shirt and pressed his lips to my breastbone. "Then I'm happy to give you what you need." He undid the second button of my shirt and placed a kiss between my breasts. I threw my head back and rocked slightly. I wanted him. Badly.

“Then I suppose...” He gripped the back of my head. “I’ll have to...” He nipped at my neck, and a short, breathless laugh escaped me. “Get you some food.”

He grabbed my waist with both hands, lifted me from his lap, and stood up.

“Huh?” I was having difficulty forming a coherent thought through the haze of my lust.

“Breakfast, my love?” One side of his mouth lifted in a half-grin.

“Wait, what now?”

I clambered to my knees. His half-grin gave way to a chuckle. He bent down, lifted me, and threw me over his shoulder in a fireman’s carry. I squealed, and he tossed me onto the bed. I bounced twice and laughed, the joy I hadn’t felt in decades bubbling through me. He leaned in and kissed me breathless before straightening back up. He adjusted himself so the bulge in his pants wasn’t quite so obvious.

“I’ll be back with a pile of bacon higher than you’ve ever dreamed.”

“Pigs would end up on the endangered species list.”

“Challenge accepted.”

After one more seductive kiss, he turned toward the door. Halting with his hand on the doorknob, he shot me a smile so sexy it made my breath hitch. I returned his grin with a mock pout.

"We're going to eat, sweetheart." His smile turned slightly wicked. "Because we'll need a lot of energy for what I have planned for you."

With that, he left, and I collapsed back onto the bed, a big, silly grin spreading across my face. I couldn't recall the last time I had felt this happy.

PART TWO

DRACO

Nothing could wipe the grin from his face. He was pretty sure he looked like he'd slept with a hanger in his mouth. Whistling a Dean Martin tune, he skipped down the stairs to get Gelli some breakfast treats.

As he neared the landing, he slowed, belatedly remembering they were in Isaiah's house. Damn. And he was not the guest of the hour right now. Rifling through his pantry was unlikely to be the fastest way back into his good graces. Double damn. Recalling his behavior at dinner left him feeling sheepish. He would make sure to apologize to him the first chance he got.

My friend? Are you happy?

"Mischief! My darling girl!" His familiar rounded the corner and froze, her eyes wide. She hastily backpedaled in shock, but he swept her up and cradled her, spinning her around as he did.

What's happening here? You're making me dizzy, she exclaimed as he stopped swinging her around and held her close.

"Mischief," he said, nuzzling her fur.

She narrowed her eyes in irritation and laid her ears against her head. Attempting to break the contact with a stiff paw to his face, she found his hold firm. Physical affection was not her thing, especially when it seemed arbitrary, like children trying to pet her, confusing her in addition to being undesirable. Despite her physical challenges, she could move pretty fast when she needed to.

Put me down immediately. I haven't bitten your face since I was a mouthy little kitten, but so help me.

He gently placed her on the floor and stroked her head. She took two stumbling steps backward and sat out of reach.

Are you going to explain yourself, or am I to launch my current plan of a vicious attack on any exposed skin I can get to before finding somewhere to hide until you are back to normal?

"Have you explored this place?"

She stood, retreated a few more paces, and sat again with an annoyed huff.

Your toes are officially in grave danger, friend. Final warning.

She glared at him.

"Would you be able to help me put together a plate of food?"

What do you take me for, Draco Winter?

She stood and stalked back toward him and his now gravely imperiled toes.

Do I look like I work here? How on earth could you dare be so brazen as to-

"She kissed me, Mischief."

She blinked her green eyes at him, and her back end wobbled before plopping to the carpet. She tilted her head, and her whiskers trembled.

She did? Truly?

He nodded. She shot to her feet and bounded at him full-tilt, hooked her little claws into his pants, and began to climb.

Pick me up! Pick me up! Pick me up! Pick me up!

"Ow, ow, ow, Mischief, okay!"

Laughing, he reached down and lifted her. She tucked her head under his chin.

I forgive you for behaving so strangely and confusing me.

He suppressed a chuckle. He didn't want to risk her thinking he was laughing at her when she was being magnanimous. No cat responds well to being laughed at, magical or mundane.

Can we go see her now?

"I was getting food for her, remember?"

Okay. You get the food, I'll go see her.

This time, he did chuckle.

"That's fine, my friend. I'll be up as soon as possible."

She trotted off, gaining momentum until she got her feet under her and could manage a good clip. Then she disappeared up the stairs.

He set off in the direction of the dining room where they'd been earlier. It seemed as logical a place as any to begin searching for either the kitchen or a person who could help him. However, it seemed his sense of direction was not as infallible as he'd hoped, and he found himself wandering a bit lost.

The place was enormous.

It turned out there were multiple dining rooms, none of which he stumbled into being the correct one. He tried to keep his impatience in check, but it was becoming difficult. All he wanted was to get back to Angelica. But he also wanted to bring her a massive plate of all her favorite things to eat.

Hell, he wanted to give her the world.

He daydreamed of her as he walked, this time with elation and gratitude versus his longtime companions, guilt and regret.

Shadows dominated the space, and the ceilings were so high they could have been supported by wood, metal, or grilled cheese sandwiches, for all he could see.

He'd just about given up hope of ever finding his way back to civilization when he heard the faint echo of a voice. He was almost certain it was Isaiah, whom he wished to apologize to anyway, so he strained his ears and set off in the direction it appeared to be coming from.

The voice grew more distinct the closer he got, and he relaxed a bit. Finally. He'd begun to feel as if he'd been trapped in a maze designed by a lunatic. Approaching a large, heavy door of polished, dark wood, his footsteps were muted by thick, ornate oriental rugs. He realized the sound he'd heard was singing. Behind the door, he could just make out an Abba song being belted out in off-key warbling. What Isaiah lacked in musical ability he more than made up for in enthusiasm.

The song halted mid-note at his knock. Draco's ears breathed a sigh of relief. Shuffling footsteps on deep carpet approached, and he stepped back as the door creaked open. Isaiah peered at him through the crack, his silhouette framed by the lighted room behind him.

"Mr. Winter," he said with surprise evident on his face.

"Good evening, Isaiah. I'm very glad to have found you, for a number of reasons."

He tilted his head in a way which reminded Draco of Mischief.

"Oh? How may I help you, Mr. Winter?"

"Draco, please."

He acknowledged that with a nod.

Draco took a deep breath.

"First and most importantly, I owe you an apology. I was way out of line this evening at dinner."

Isaiah's shoulders visibly relaxed at my apology. Then he shrugged.

"Forgiven. I understand this is a time of great stress for you three."

Understatement, Draco thought.

"Yes, it is. And second?"

"I'm, well," He rubbed the back of his neck, "a bit lost."

His lips twitched, and Draco got the distinct impression Isaiah was holding back laughter.

"Yes, this place can be a little labyrinthine. Come," he stepped into the hallway, shutting the door behind him, "I'll get you back to your room."

"Actually, if it's not too much of an imposition, I was hoping to put together a plate of food for Angelica."

His eyebrows shot up to his hairline.

"Is that so?" He clasped his hands in delight. "Interesting. Wonderful, really. You two came together earlier than I'd anticipated. No matter," he continued, oblivious to Draco's stunned expression, "let's get you to the kitchen, then."

He strode past Draco, already halfway down the hall before Draco shook off his surprise and began to follow, jogging a little to catch up. Isaiah was practically skipping, smiling and humming to himself. His sudden change in demeanor was a little dizzying from earlier.

Draco! Mischief's voice cried in his head. A knot of dread settled in his stomach at the distress in her voice. He stopped abruptly, and Isaiah turned to him, the smile melting from his face.

Draco, come quick! It's Angelica! Something's wrong!

His blood turned to ice in his veins. Brows furrowed, Isaiah's expression changed from inquisitive to concerned.

I'm coming, Mischief. Just hang on, okay?

Okay. Please hurry, Draco she said urgently.

He nodded silently to her, though she couldn't see him.

On my way, little one.

Aloud, he said, "We need to get to Angelica's room. Now. Something's wrong."

Isaiah nodded wordlessly and took off, his robe billowing behind him.

~

Angelica lay on her back on the bed right where he had left her. Mischief's head turned toward him when the door opened, her eyes panicked. She was perched on Angelica's stomach, kneading her paws into her chest.

I came up to be with her, and I thought she had fallen asleep, so I started singing loudly like I used to when I was hungry in the mornings and you two wouldn't get out of bed. But she hasn't woken up, Draco! She won't wake up!

He crossed to the bed in two swift strides and dropped to his knees. He cupped her face in his hands and patted her cheek gently.

"Gelli, wake up. Come on, wake up, sweetheart."

No response. Her eyes were closed, but it was clear she was not asleep. She was unconscious. There was nothing restful about her expression. Her eyes were closed too tightly, and her mouth was pinched into a tense line. Her forehead was creased, a furrow between her eyebrows wrinkling her beautiful skin. Her newly auburn hair spilled over the comforter, the shock of white a shade or two darker than the fabric. Panic rattled its cage in Draco's chest.

"Angelica, open your eyes, baby!"

"What's going on?" Kaden strode into the room, rubbing his wet hair with a towel.

"Angelica has fallen unconscious. We don't know why or how. Mischief just found her like this."

Kaden dropped the damp towel to the floor and jogged a few steps to the bed. He leaned one knee beside Draco on the mattress and picked up her wrist.

"Her pulse is elevated. She also feels warm to the touch. Do you think she's fallen ill?"

"She was fine ten minutes ago when I left her room."

"Shit. Do you think... Wait, when you left her room? What's going on?"

"Later." Draco removed his hand from her face and fetched a pillow, gently tucking it under her head.

"This reeks of magic," Isaiah stated, pushing off from the wall near the door. Draco had almost forgotten he was there.

"Excuse me," he said, gesturing for them to clear a spot for him next to the bed. Draco reluctantly moved away. Isaiah sank to his knees and reached out, his hands hovering just above her skin. He froze.

"Not just magic. Her magic."

“Whose magic? What the hell is going on?” The distress in Kaden’s voice mirrored Draco’s panic.

Isaiah stood up, fisting his hands at his sides.

“The Crone.”

Party Like It's 1699

Irate didn't even begin to describe Damien's state of mind. He couldn't decide whom he loathed more at this moment—himself or The Crone.

Once the initial part of the curse, his sole focus had been completed, all he'd wanted to do was get to work. All he'd wanted was The Crone to finalize the last step, enabling him to infiltrate the little witch's mindscape and annihilate her. But instead of moving forward, he was now ensnared by the fine print—a rookie mistake.

It's the first lesson in drafting a satisfactory demonic contract. Be sure to include the fine print. He had failed to do just that.

He hadn't stipulated a deadline for The Crone's services, and now he was at her mercy, a grave oversight on his part. His mind oscillated between admiring her talent and therefore the value of allowing her to continue breathing and wanting to rip her head from her body with his own hands in a frenzy of rage and bloodlust.

At some point during the evening she had fixated on him as her new toy. A prize she'd won. He'd initially intended to use it to his advantage. Get a few more services from her, gratis. Possibly lift her by her thighs and fuck her against the wall. Instead, he now found himself chained to a chair, seething with anger, compelled to play a game he hadn't written the rules for. So, instead of swearing and furiously rattling the iron restraints she had conjured around him, he was forced to feign civility and smile until she decided to finish helping him.

And when that moment came, he would end her.

She sashayed in, humming a tune he'd heard on the radio over the past few weeks, "Possession" by Sarah something-or-other. He struggled to suppress an eye roll.

Once again, she had changed her outfit, this time wearing a gossamer, transparent dress which hugged her curves and left nothing to the imagination. Adorning her ankle was a delicate silver anklet adorned with tiny bells.

He fought the urge to rip it off and shove it down her throat.

He flashed her his most charming smile. Frustrated and angry as he was, he couldn't lose sight of the fact she was still a psychopath with a short fuse and an unpredictable trigger finger. It would be a pity to come this far only to be annihilated by a madwoman.

She stepped over to him , she swung a leg over his waist, straddling him. Despite his seething at her and the situation, his visceral reaction to her body astride him stiffened his traitorous cock.

He could discern her erect nipples through the sheer fabric as she leaned in, pressing her breasts against his bare chest. Lowering her head, her soft tongue flicked out to taste his throat, her teeth grazing his skin lightly. She moaned softly. He gave a mental shrug. May as well enjoy the situation as much as he could.

The chains had enough length he could reach her waist, so he grabbed her hips and began massaging them with his thumbs. She pressed her core harder against him and his cock wept. He had grown almost painfully hard.

"Mmmm, Damien..."

The sound of his name on her lips stirred something deep within him. The last time he felt anything remotely similar was for Ameline in the late 17th century, before she was murdered by zealots. Intriguing.

She removed his hands from her waist, and he snapped back to the present. With half-lidded eyes, he watched as she slithered down his chest. Just as she reached to unbuckle him, she paused and gazed up at him with a look of intense desire and lust, causing his arousal to intensify.

"I could relieve some of that pressure for you, darling. Would you like me to do that?" she purred.

He nodded.

"You need to say it, Damien," she insisted.

"What is it I need to say, love?" he managed, his throat dry, his traitorous dick. .

"That you want to fuck me."

A muscle twitched in his jaw.

"Yes, I want to fuck you," he replied.

She gave him a small smile.

"Say you want me."

"I want you."

"Tell me I'm beautiful."

"You are. Incredibly beautiful."

"Tell me you want to come. Tell me you want to release inside me."

"I want to explode deep inside you."

"Tell me you're mine. My slave forever."

He saw red. Enough of this game. With a surge of rage, he snapped the chains and tore the cuffs from his wrists. He shoved her to the floor and stood over her.

"That's enough, witch."

She gaped at him.

"I may want you, but I am nobody's slave."

“How did you break those chains? They were iron and spelled. You shouldn’t have been able to do that.” She crab-walked backward as he stalked toward her.

"Because I'm Damien." His voice dropped and octave and spoke with the sound of a dozen demons.

"We are Legion."

Fire licked at his fingertips. Her eyes widened as she continued to scramble back. He pursued her until her back met the wall. Growling, flames climbed higher. Backed up against the wall, The Crone trembled.

Then her posture relaxed and she smirked at him. He hesitated.

"Fool. You really think a little pyrotechnics and the voices of hell scare me?" She stood, brushing off her hands, anger blazing in her gaze. Her dress vanished, replaced by black leather from head to toe. A cat o' nine tails dangled from her left hand, an obsidian blade glinting in her right. Cracking her whip against the floor, she strode toward him like a vengeful goddess. Now it was his turn to gape.

Damien snuffed out the flames, the tension in his pose vanished. His cock, softened by his earlier outrage, began to swell again. He suddenly ached for her. She looked stunning and fearsome in all her wrathful glory; he wanted to worship every inch of her with his tongue. Flashing her a predatory grin, he approached. She stopped and regarded him quizzically.

Closing the gap between them, he grabbed her roughly, his hand clutching her hair. Yanking her head back, he kissed her fiercely. She dropped her weapons, wrapping her arms around his neck. He swallowed her moans and growled a warning when she attempted to move. She gouged his back with her fingernails, drawing blood. Finally, he allowed her to break the kiss.

"Damien…?"

"I want you."

A corner of her mouth twitched. He smirked in response.

"You are the sexiest thing I've ever seen."

Her smile widened, desire lighting up her eyes.

"I'm going to bury myself deep inside you. I'll make you come so hard you'll forget your own name."

She licked her lips.

"I'm going to pleasure you until all you can remember is my name and my touch. I'm going to make you scream."

Her lips parted, breath quickening. “I'll never be your slave, but you'll beg to belong to me.”

A talon extended from his fingertip as he ran it down the front of her clothing, tearing it apart. Lifting her by the thighs, he backed her up against the wall as he had envisioned. With one hand holding her up, he unbuckled his belt with the other. His cock sprang free, and he returned his hand to her thigh, spreading her legs wider. Thrusting into her, she threw her head back with a keening cry. He began pounding into her with ferocity, matched by her, dragging her fingernails down his chest and back until his skin was shredded into ribbons.

As her inner walls spasmed and clenched around him, he increased the pace, slamming her against the wall until one wood panel cracked. His lips found her neck, and he struck, sinking his teeth into her throat. Her muscles clamped around him, and she screamed. Releasing her neck as his balls tightened, climax imminent, he thrust three more times and came with a roar.

Her pulsing sex milked every last drop from him before he glanced around the room. Iridescent filaments of red and black filled the space. Light pulsed in the corners in sync with their heartbeats. A wicked smile spread across his face.

It was as he suspected. When she finally brought her gaze down from the ceiling and looked at him, flames flickered in her eyes. He knew from her astonished expression his eyes held the same.

She was his mate. Together, they would be unstoppable.

“Damien.”

“Yes.”

“Does this mean what I think it means?”

“Yes.”

She pursed her lips and glanced around the room, a crease forming between her eyebrows as she regarded him warily.

“Not some demon illusion?”

“If it is, it’s not mine.”

“Oh.” She pressed her lips together. “What are we going to do about it?”

He couldn’t suppress the slightly manic smile that crept onto his face.

"We're going to rule the fucking world, my dear."

"Hm." She smiled at him, her eyes sparkling with a hint of cruelty hidden beneath the surface. "Then we'd best get to work."

He dipped his head to her chest and bit her nipple hard enough to draw blood. She gasped, and the scent of her arousal enveloped him.

"Later," he replied with a bloodied smirk. "We can leave the little witch on ice for now. The next forty-eight hours will be dedicated to consummating our unholy mating." He bit her other nipple and thrust two fingers into her. She moaned and began to writhe against him.

"More," she panted desperately, the single word sending a jolt straight to his groin.

"In good time," he said, grinning up at her.

She gripped his hair tightly and pulled. He sucked on her tongue.

"Now," he said, pulling away to meet her blazing eyes, "where did you drop that whip?"

~

Angelica lay against the pillows. It had been almost twelve hours with no change. Still the mild fever, the elevated heart rate, and expressions which ranged from mild tension to abject fear. It seemed as if she were caught in a nightmare, and Draco was desperate to wake her from it. He just didn't know how.

He felt powerless. Impotent.

Isaiah had sequestered himself in his library to research. The fact that the Ancient did not know what was going on and, more importantly, how to fix it was alarming, to say the least. At least he'd decided to help and not shove them out his front door.

In fact, Isaiah was now officially on board with their mission. Since it had happened in his home, he took it as a personal attack and was seriously pissed about it. After discovering her in that state, he'd tried multiple things to get a response from her, and when those failed, he nearly put his fist through one of the walls before heading straight for his library ten hours ago.

Isaiah clearly knew of The Crone to be able to sense her magical signature so quickly. Draco suspected they had a history, but Isaiah was evasive when asked about it.

He sighed. He was so weary. Kaden tried multiple times to get him to leave her room, but there was no amount of cajoling that could tear him away from her now. He couldn't bear the thought of her suffering alone, even if she didn't know he was there. He expected her to be terrified and confused when she woke, and he wanted to be there for her. Hold her until she felt safe again.

Damn it.

"Please wake up, baby," he whispered for what felt like the thousandth time.

He briefly wondered when he last ate. Kaden had brought food several times, but Draco had turned him away. His stomach, twisted with worry, rendered his appetite inert. As if summoned, Kaden walked in again with a plate of food.

"Hey, man," Kaden greeted him without breaking his gaze from Gelli's face.

"Hey," Draco responded, hearing the sound of a plate being set down.

"Look, Draco, you have to eat."

"I can't."

"Well, you must. And take a shower. Maybe get some rest."

Draco heard Kaden sigh and turned to face him. Worry lined Kaden's face, and dark circles hung under his eyes. Draco cocked his head at him.

"Maybe you should take your own advice, man. You look awful."

"Thanks."

"Any time."

"The point is, nothing's changed in twelve hours, and a shower, some food, and some rest will help you feel better."

Draco gave him a measured look. "Kaden, I spent twenty-seven years eating, showering, and sleeping without her. I can assure you, those things will not help me feel better."

"But she's not gone. And she will still be with you even if you leave her room for twenty minutes."

He opened his mouth to argue, then shut it. Kaden had a point.

"Hey," Kaden said, crossing the room to sit in the chair next to Draco, "Correct me if I'm way off here, but I think I understand what you're afraid of."

Draco raised an eyebrow at him. Kaden rolled his eyes and lifted his hands in surrender.

"Whatever. What you're worried about. You just got her back and you think she'll suddenly disappear from your life again if you take your eyes off of her." He clapped him on the shoulder. "But even if that were a reasonable concern, which I assure you it is not, this situation has nothing to do with the two of you. It is an attack on her and it concerns all of us."

He waited while Draco mulled over his words.

"I think you may be right."

"I usually am."

He stood with a small smile and turned to leave. Draco's expression must have given away his thoughts, because Kaden stopped.

"What's on your mind?"

"I was just wondering whether you are descended from a line of empaths, actually. Your perceptiveness is spooky accurate."

He shrugged. "I don't know. Abandoned as a pup. Wouldn't be surprised if I were, though."

Draco nodded. "And Kaden?"

"Yes?"

"Thank you for being Angelica's best friend." Guilt dragged his gaze to the floor. "I know I don't exaggerate when I say you saved her life."

From me. When I broke her heart.

Draco hesitated before looking back up at him. Kaden stood with his hand resting on the doorknob, regarding him for a moment before speaking again.

"You're welcome," he replied, leaving the room with a wave. "Your plate is on the dresser. You should get to it before it gets any colder," he called over his shoulder before shutting the door. Draco stood and retrieved his plate before settling back in his chair.

After he showered, he put on some clean sweats and climbed into bed next to Gelli. Her warm skin felt divine against his slightly damp body. He draped an arm around her and pulled her close, breathing in her scent deep. She'd always smelled so good. Like lemongrass and peppermint. He gave her shoulder a kiss.

"We're going to figure this out, sweetheart. I promise."

He slept with her wrapped in his arms all night.

Four days later, everything went sideways.

Something Wicked This Way Comes

Damien didn't know whether to send his thanks up, down, or sideways. That The Crone should be his mate was better than a best-case scenario. She matched him in every way. Her epic lust for sex, for cruelty, for blood, and for power was rivaled only by his own. It was perfect.

The final test had been revealing his true form to her. Had she become frightened, it would have been clear they were not mated. Succumbing to madness and losing her talents would have been a travesty. But she surprised him once again.

He'd watched her carefully for any negative reactions as he allowed his real form to manifest, but instead of horror or insanity, even indifference, he saw something else entirely.

Lust.

They spent days consummating the bond over and over. And over again. They paused only for hydration and sustenance. She even went into a nearby town and snagged him a sweet little morsel of a teenaged girl. Calliope called it a 'bonding present.'

In searching out something for her and her in-house apothecary, he discovered another reason for celebration. He was getting stronger. Strong enough, in fact, that he could once again jump through the Void.

The Cage

Anyone who has lived by the ocean can tell you it's easy to discern the tides just by listening. The sound of high tide is soft, with gentle waves. Low tide sounds like slaps on the sand, accompanied by discontented rumbles.

It was the sound of the tide which gave it away. It never changed.

The surroundings were lavish. Beautiful. Fine silks and linens draped luxuriously. The bed was gloriously soft, meals delivered were delicious, and the bath put the Romans to shame. Yet, the light from outside only varied from twilight to night. Time washed over me without regard, and I found myself withdrawing further and further into myself. Whether I'd been there a week or a month became impossible to determine. Anytime I felt even the slightest glimmer of hope, the doors of my prison clanged louder than church bells.

"Your meal, Miss."

I nodded as he set my fare on a nearby table. In the initial days of my capture, I had tried to engage the servants in conversation, only to end up wanting to bang my head against the wall in frustration. They were not here for me. I now knew they weren't even real. Though, at this point, 'real' was becoming a concept too slippery to hold onto.

I knew this room was a figment. Wasn't real.

I knew the servants weren't real.

I knew the palatial gardens and the fountains weren't real.

I knew love stories weren't real.

I knew darkness. And the feel of my fingers as I wrung my hands. The pressure behind my eyes when I refused to cry. I had my suspicions as to whom my jailer was, but when I tried to focus on the thought, my brain became fuzzy and languid and the name would slink away.

Today, food turned to sand in my mouth. Anxiety nudged the recesses of my mind. There was something I needed to remember...

The door swung open, and a man strode in.

No, not a man. Something else.

“You look familiar…” My head felt fuzzy again. I screwed my eyes shut and pinched the bridge of my nose, trying to shake my thoughts.

He hummed in agreement and approached.

“Open your eyes, Angelica.”

I complied and looked up. His eyes were pools of swirling black and violet. The longer he held my gaze, the clearer my thoughts became. Fear seeped into my limbs, my hands trembling.

“You’re…” I started.

“Malachi,” he said, extending his hand for me to shake. His smile chilled my blood.

“And we have much to discuss.”

~

Sometime in the night, Draco had snuggled up to Angelica, his face buried in her hair. Surrounded by her unique scent, in the fugue state between waking and sleep, contentment filled him. He felt the morning sunlight on his back and heard the ambient sounds of the waking forest surrounding Isaiah's house.

Suddenly, reality crashed down on him. The smile melted from his face, and despair and nausea cramped his body.

He was tempted to retreat back into sleep until he realized what had woken him in the first place. Next to him, Angelica was mumbling something. At first, it seemed incoherent, but after listening for a bit, her words began to take on the cadence of a prayer. Galvanized by the prospect of progress, he slipped his arm from underneath her and got up to seek out Isaiah.

Draco found Isaiah in his library, surrounded by books and scrolls, poring over a manuscript. Though the door was open, Draco knocked on the doorjamb to announce his presence. Isaiah looked up at him with bleary, bloodshot eyes.

"Did you get any sleep?"

"I caught a nap sometime in the early morning." He sighed heavily and scrubbed a hand over his face. "What can I help you with, Draco Winter?"

"There's been a development. I need you to come with me."

He raised his eyebrows and reverently set aside the tome he'd been examining.

"After you."

Draco walked back to Angelica with Isaiah in tow. They were about halfway down the hallway from the bedroom when Isaiah picked up his pace and passed Draco. He jogged to the bed, his robes flying behind him. He bent at the waist and hovered a hand over Angelica's crown chakra.

"I thought it sounded like a prayer. Are you able to translate what she's saying?" Draco asked. Isaiah glanced over his shoulder with a nod and a small smile.

"It would seem Angelica is stronger than we could have hoped," Isaiah replied.

"What do you mean?"

He paused his hand briefly over her throat before moving to her chest. "It is not a prayer, though I understand why you would think so. No, it is a protection spell. A very old spell. And…" He furrowed his brow. "If I'm not mistaken, it's Enochian."

"Enochian? As in the language of the angels?"

“The same,” Isaiah said with no small amount of awe.

Draco’s knees went a little weak, and he leaned against the wall for support. “How on earth would she be able to wield a spell like that?”

“Not on earth.”

Draco huffed a frustrated breath.

“Figure of speech, old man. You know what I meant.”

“As I said, she’s stronger than we’d thought.” Isaiah nodded brusquely. “You were right to get me. This requires more research, but at least now I have a clue as to where to begin.”

With that, Isaiah hustled out of the room and back to his library. Draco dropped to his knees at Angelica’s bedside, clasping his hands in front of him on the bed. She may not be praying, but he sure as hell was.

Later, Draco found Kaden pacing in his room with Angelica’s Tarot deck clutched in one hand.

“Tell me you have some good news,” Kaden said before Draco even entered the room.

“I’m not certain whether it’s good, but it’s news.”

“Tell me,” Kaden replied in a tone sharper than Draco had ever heard from him.

“I awoke this morning to Angelica mumbling something I thought was some kind of prayer. I grabbed Isaiah and he said she’s spell-casting.”

“Why would you need Isaiah to tell you that? Surely you’d recognize a spell when you heard one.”

“He said it’s Enochian.”

Kaden abruptly stopped pacing.

“I’m sorry, but it sounded like you just said she’s wielding an angelic spell.”

“That’s what I said.”

“Holy hell.”

“Apt.”

Kaden flopped into a chair and propped his arm on an intricately carved writing desk supported by legs made to look like griffin’s claws. Draco walked over and his eyes were snagged by a small scene near Kaden’s right elbow.

“This is a beautiful desk.”

Kaden nodded with a shrug.

"Kaden, how closely have you looked at these carvings?"

He looked at Draco with a mixture of puzzlement and irritation.

"Seriously, man? Pepper is comatose and you want to talk about furniture?"

Draco shoved Kaden's arm from the desk, spun his chair around, and pointed to the section which had caught his attention. Kaden glared at him before turning his attention to the carvings. Then he stilled.

"Is it just me, or are these images...moving?"

"Yes. And if I'm not mistaken, I would say those are soldiers readying for battle."

"Holy shit." Kaden sat back in his chair.

"Amen."

~

"I need a sacrifice," The Crone announced from the kitchen.

"Of which type?" Damien asked, toying with one of the poppets she had lying around.

After a quick phone call Downstairs, he confirmed his suspicions that his bond with Calliope was hellsent. He wasn't deluded enough to consider it a reward for anything but a strategic move, possibly orchestrated by his father or someone even higher up the food chain. The idea made him smile as he slowly tore the limbs off the doll.

Siren lay by the fire, one eye trained on him. Despite the bond, or perhaps because of it, the cat would probably never like him. He was fine with that. He didn't care what an overgrown housecat thought of him.

As if hearing his thoughts, Siren let out a low growl and flicked her tail. Damien got up to join Calliope in the kitchen. He may not care what the panther thought of him, but he wasn't stupid. The Crone's familiar was still exceptionally dangerous and formidable. Forgetting that would be a grave mistake.

Calliope turned away from her cauldron when he entered the kitchen.

"Stop antagonizing Siren. She doesn't appreciate it."

"I wasn't."

The lights dimmed.

"I don't appreciate it," she replied in an icy tone, a reminder that forgetting how dangerous and formidable she was could also be a grave mistake. "And I told you not to lie to me."

He narrowed his eyes.

"Dare you threaten me, witch?"

Frost crept up the windows and the lights flickered aggressively. He could hear Siren's growl from the sitting room, adding to The Crone's displeasure.

"Indeed, I do. Being mated to me does not guarantee your survival. I will not hesitate to kill you if I see an advantage to it." Her eyes filled with malice. "Or should you give me reason to seek a pound of flesh from you."

He smirked.

"Is that the sacrifice to which you referred? It would render this project somewhat moot should that be the case, no?"

She turned back to her work without answering him. He knew nothing she'd said was untrue, as he felt precisely the same. They were, after all, two sides of the same coin. But it was still her game. For now.

Despite adopting a cavalier attitude, Damien felt a tickle of unease. More that she would choose to stop the spell indefinitely than fear of bodily harm. She was contractually bound to finish the spell, but again, he had failed to include the fine print designating when it needed to be finished. Impotent rage burned within him. Instead of unleashing his ire, he strode out of the house without a word. He could have sworn he saw a smug smile on Siren's face as he passed by her.

Training Grounds

Malachi kept hold of my hand and pulled me to my feet.

“This may hurt. Try not to scream.”

He held one hand in mine with the other pressed against my chest. At first, it felt pleasantly warm. Then it got hot. Soon, it felt like he was setting my skin on fire.

“Stop! Stop! What are you doing?” I cried out, trying to pull away from the searing pain.

“I’m breaking you out, Angelica,” he said, holding me firmly. “Look around.”

Despite feeling as if his hand were melting a hole in my chest, I breathed through it and tried to do as he asked. Looking around, I began to pick up on subtle differences. The cutlery was now silver instead of gold. The bedding was green instead of blue. The sound of the waves had vanished.

"What the...?"

I struggled to understand what was happening, feeling as if I was reaching the limit of what I could handle. The striped silk wallpaper began to peel away in long strips, and the walls dissolved into puddles of plaster. Fear gripped me, and tears threatened to fall.

"What is happening?" My voice sounded weak and tremulous. I cringed at how pathetic I sounded.

Malachi huffed an impatient breath, flicking his fire-brimmed eyes to consider me for the briefest moment.

"Almost there." He rolled his eyes skyward. "You're doing just great."

The insincerity in his tone and my annoyance at his gall edged out my fear aside. He captured my gaze once more.

"Okay. Now, this may hurt."

"Are you messing with me right now? What has this been so far?"

"The warm-up."

His eyes flashed a glowing violet, and the pressure on my chest increased. Suddenly, there was a boom like striking lightning, and I was flung across the room. I sailed effortlessly through the now transparent walls, landing gracelessly on a patio of some kind of decorative stone. The impact knocked the wind out of me, and I felt a rib crack. Wheezing, I tried to sit up.

"You... asshole," I gasped, throwing him the most withering glare I could muster.

"Oh, gee, princess. I thought you knew Enochian spells were powerful," he said, strolling over to help me up. That made me pause.

"As in angel spells?"

"And the prize goes to the imprisoned damsel in distress."

"Yeah, you can go ahead and start explaining all of this now."

He quirked an eyebrow and gave me a wry look.

"You really have no idea?" He shook his head in disbelief. "Fine. Come with me."

~

Calliope felt the prison crumbling. The girl would soon be awake. She rocked in her chair and petted Siren, wracking her brain to come up with a solution. Damien would be back eventually, and she would need to tell him. If she had just finished the spell in time, she wouldn't be in this situation. She had wanted to keep Damien on the hook, but now they had a problem. The door opened behind her, and she closed her eyes.

"Honey, I'm home."

~

Malachi grabbed my hand again and dragged me behind him at a furious pace. I began to feel as if I were a toddler being hauled away for misbehaving at a carnival. I stumbled and bounced along until I finally dug in my heels.

"Wait! Enough! Tell me what the hell is going on!" I jerked my arm free from his tight grip. He stopped abruptly and regarded me with unnatural stillness.

"We don't have time for this."

"For what? That's my point!"

He sighed heavily.

"Do I seem familiar to you, Gatekeeper?"

"When you first came into the room, I thought you were Damien. Wait. Gatekeeper?"

"Damien is my brother."

My face went cold, and I felt the beginnings of a panic attack. I took a few stumbling steps back and shook out my hands, readying my battle magic. He watched me with a bemused expression.

"Simmer down, kitten. I pose no threat to you."

"Bullshit. You're a demon."

He tilted his head with a small smile.

"Not precisely. Do you know many demons who can wield Enochian magic? No? That's because they can't."

Overwhelmed by the hits which kept coming and the events of the day, my knees buckled. He made no move to come closer as I sat perfunctorily in the dirt. I peered up at him through a curtain of chestnut and white hair. He must have read all the questions on my face because he blew out a resigned sigh.

Waving his right hand, the ground split in a spray of dirt, and a tree stump emerged.

"I would have preferred to go over everything in more comfortable environs, but you seem determined to frustrate those efforts."

I shrugged, lacking the energy to argue.

"You are special, Angel. Didn't you ever wonder how a black-hearted witch and an ancient vampire could create life? The short answer is, they can't."

I nodded, having wondered but never brave enough to uncover what I suspected were unpleasant truths.

"There is a balance to everything, Angel. I say Damien is my brother, but it isn't strictly true. We were made to balance each other. Keep each other in check. I am a demon, but I am not."

"Clear as mud. Please continue."

He rolled his eyes but complied.

"Where he is dark, I am light. Where he is positive, I am negative."

"I was unaware he had positive qualities."

He tilted his head side-to-side.

"Depends on perspective."

"Okay." I shifted to move a rock poking me in the butt.

"The point is, I want him back in his cage."

My eyebrows shot up. This was unexpected.

"And you're the only one who can do it, Gatekeeper."

"Okay, that's the second time you've called me that. What does it mean?"

He closed his eyes and sighed.

"I can't explain here. It is not safe to speak freely of such matters. We have to leave."

This time, I didn't complain as he dragged me along like a doll tied to a string.

We traveled for another mile or so before Malachi stopped and dropped my hand. I bent over at the waist, hands on my knees, taking in great gulps of air.

"You are not in great shape, Angel."

I flashed him my middle finger in lieu of trying to speak. He ignored me and began drawing runes in the air about him.

"What…?"

"Quiet. No questions."

I snapped my mouth closed and watched him work. His technique was something to behold. He was grace and beauty in motion. I stood transfixed and watched him as my breathing returned to normal. He flicked his wrists and closed the spell. Nothing happened.

"Is something supposed to?"

"Patience."

"Not my strong suit…" I mumbled to my feet.

After a time, tremors began to reverberate through the soles of my shoes. Nearby trees whistled and swayed as a strong breeze picked up. The air felt galvanized, and the hair on my arms stood to attention. Soon, the tell-tale ripple preceding a portal opening shimmered in a chaparral of trees beside us.

"Oh, hell no." I shook my head. "I'm not hopping into a random portal with a demon I just met. Thanks, but no thanks."

"You have another option, then? I suppose staying here qualifies."

He straightened his jacket and strode slowly toward the materializing portal. He glanced over his shoulder, poised to enter the undulating haze of color and light. My shoulders slumped.

"Fine. Just don't eat me, okay?"

He shook his head. “Disgusting,” he muttered before stepping through. I felt oddly insulted.

“Damien always wanted to eat me,” I said to the retreating form once I was sure he couldn’t hear me.

I trepidatiously approached the portal. I’d not had a lot of luck traveling through them. Historically, they’d made me violently ill. But, considering my other option was staying in this no-place, I was willing to see what Malachi had to offer. Maybe I’d puke on his shoes as a bonus.

I stepped in and was immediately sucked into a vortex of whirling color and form. I felt stretched beyond my limits and then compressed to the point of pain. I fucking hated portals.

“Damnit, Gelli.” Draco kissed the spot behind her left ear, her hair tickling his nose. Isaiah had sequestered himself in his library, and Kaden couldn’t seem to stay still. When he wasn’t working out, he was running. When he wasn’t running, he was pacing and chewing his nails.

She stopped muttering in Enochian and was quiet. she took a deep breath and her eyelids began to flutter.

“Kaden! Isaiah! Get in here!” Draco shouted.

There was a crash down the hall, followed by muffled swearing. He heard Kaden's rapid footsteps as he jogged toward Angelica's room.

"What's up," he panted, hand on the door jamb.

"Come here." Draco waved him over and scooted back so he could get a better look. "Is it just me, or does it seem she's waking up?"

The Half-Flayed Man

Damien strode into The Crone's lair, immediately sensing something was wrong when Siren sidled up to him and bumped his side with something akin to affection.

"What's the meaning of this, Calliope?"

"Siren can't say hello to you, demon?"

"She hates me. Tell me what is going on," he demanded, feeling increasingly unsettled.

The Crone sighed and stood.

"The prison has been breached."

Rage washed over Damien like a molten shower.

"Repeat that," he growled.

She set her jaw. "You do not order me around."

"In this case, it's justified. What. Happened?"

She shook her head and strode to her kitchen. Damien followed small flames springing up in his wake. The smell of brimstone and singed wood wafted after him. She approached her cauldron and waved a hand over it.

"See for yourself."

Damien peered over the edge at two figures running through a forest.

"Is that…?"

"Yes. Your brother."

Damien felt a fury unlike any he'd ever experienced. Someone was going to pay for this.

"You." Growls reverberated through his body as he turned to face The Crone.

"Not I. Your brother." She took two steps back, and the cat o'nine tails unfurled in her right hand.

"You!" he raged, sweeping the cauldron from the table. It flew across the room and crashed into her apothecary table. Bottles and vials shattered, mixing tinctures and potions intermingled, and an acrid smell filled the room. Smoke billowed, and Siren roared in outrage.

"Why did we sit on her? Why didn't you finish the spell, witch?"

Damien punched the kitchen window, shattering it into millions of shards of leaded glass. He whipped his hand around so his blood hit her, sizzling and burning holes into her dress. She looked down and snarled. With a violent wave of her arms, the dress disappeared, and she was once again clad in head-to-toe leather.

"You want to play with me, child?" She cracked the whip and split the tops of his shoes. She reared back for another swing, but Damien caught the middle of the whip, yanking her closer.

Siren launched herself through the air, landing all 700 pounds of furious, taloned cat on his back. He shouted in pain and rage. Siren roared and struck at his neck, but he was already shifting to his true form, and she got a mouthful of impenetrable demon scales.

His form peeled back, and flames began to lick the sides of Siren's face. She howled and retreated. The Crone's face twisted in rage and disgust.

"How dare you? You come to me, I allow you into my home, and you attack Siren? Inexcusable!"

A razor-sharp metal spear, shining like polished glass, appeared in her left hand. She aimed and threw it, slicing through the air and embedding it eight inches into Damien's chest. Ichor poured from the wound as he yanked it from his flesh, the spear smoking and sizzling on the floor where it fell.

The Crone screamed in frustration and produced an executioner's ax from the aether. She charged at him, screaming like a banshee, intent on separating his head from his body. Siren coiled like a viper, ready to back her mistress.

She went high, and the cat struck low, knocking Damien to the floor. The blade whistled as it descended. Damien knew a blow like this wouldn't kill him, but it would hurt like hell and possibly take him out of the fight, sending him back downstairs to recuperate.

He couldn't have that. He leaned his head forward so the blade struck and embedded itself into his left horn with a cringe-worthy squeak. The Crone yanked on the ax handle with all her might, her face a mask of rage. Realizing she couldn't pull it free, she pushed instead, and a sickening crack filled the room as half of his horn hit the floor beside him.

Damien roared and manifested fully.

"Enough of this!"

He backhanded Siren, who flew across the room, halting her progress by digging her claws into the hardwood floors. She snarled at him, readying for another attack. Damien stood. The Crone faced him and stamped her foot. The walls flexed, and the glass in the windows groaned. She threw out her hand, sending a spell into his chest like a Mack truck.

She sent two more in rapid succession, each one pushing him back several feet. She threw one more, wrapping him in chains and barbed wire, sending him falling back into a chair. He flexed and broke the chains, the barbed wire dragging down his body, taking strips of flesh with it. He stood a half-flayed creature.

"Sit down!" The Crone screamed, pushing him once again into the chair.

"No! You owe me an explanation!"

He tried to stand again, but the barbed wire cutting into his wounds only dug deeper. The sensation had him panting and his cock hardening. She strode over and punched him hard in the face before straddling him and devouring him with a kiss.

~

Finally, we stopped running. I doubled over, hands on my knees, gulping air. Malachi rolled his eyes at me. Again.

"You know, for an angelic being, you sure are impatient," I tossed his way between breaths.

"Yes, well. For a creation of the Divine, you are surprisingly lacking in stamina."

"Okay," I started, getting my breathing under control, "is it time to explain what the hell is going on yet? Or do we have a number of steps we need to achieve on your FitBit before you divulge?"

"What in The Name is a FitBit?"

"Never mind. Can we talk now?"

He nodded and waved his hands, tearing a gash in the air beside him to reveal another portal.

"Oh, boogers."

He paused and looked back at me, seemingly debating whether to ask what 'boogers' were. Deciding against more delay, he stepped through the portal and vanished. I heaved a great, long-suffering sigh and followed him.

The portal spat us out into a sumptuous library. Red velvet and heavy, carved furniture filled the space, absorbing the harsher lighting elements from the sconces lining the walls and creating a warm and inviting room. The walls were filled floor to ceiling with books and artwork. I started getting a little suspicious about where we had ended up.

"Is this Heaven?" I blurted out before dry-heaving into a nearby trash bin.

Malachi had the decency to look a little sympathetic before his face returned to resting bitch.

"Not precisely. This is my space. I've brought you here because it is the only place in the universe where I can be sure we will not be overheard."

I nodded and rocked back onto my heels. He sighed and stepped toward me, hand outstretched.

"Come on. I'll get you a tonic, and we can talk."

The shock on my face must have been something to behold because a hint of a smile graced Malachi's lips.

"I do have the capacity for compassion, you know. Remember?" He pointed to himself. "The being designed to balance Damien?"

He pulled me to my feet and led me to a wonderfully overstuffed chair. I sank in with my eyes closed and a grateful sigh. A chilled glass brushed my hand, leaving condensation along my fingertips. I took a cautious sip before drinking in greedy gulps. It was the most delicious and refreshing thing I'd ever consumed.

"Thank you. That was amazing. And I feel worlds better."

"I'm pleased."

Awkwardness threatened to invade the room, so I returned to business.

"So. 'Gatekeeper?'"

"Yes."

"I promise I won't interrupt. Please tell me what's going on."

"Yes." He nodded. "You are a special being, Angelica."

He then proceeded to talk for what seemed like hours. It went a little something like this…

Gatekeeper

Millennia ago, before the Earth was what we recognize today, Demons and Angels moved freely across The Veil. At first, all was copacetic. Demons supplemented their energies with the creatures roaming the planet, and Angels replenished those animals. Cheques and Balances. Things started going off the rails when people and their souls hit the scene. Demons went apeshit for those things.

Suddenly, our little blue planet was overwhelmed with low-frequency and lower-level demonic creatures. Infant and infirm mortality rates skyrocketed. When bigger demons visited, plagues happened. Yes, much can be attributed to the appalling lack of hygiene and general ignorance of modern medicine in our earlier days, but those weren't only to blame.

Angels were overworked and understaffed. So, the Architect created the role of 'Gatekeeper.' A neutral being with both demonic and angelic powers and the ability to banish either. One who could sense imbalances on the planet and rectify them as needed.

I had tried to stay quiet and listen, but at this point in the story, one question was just too big to keep in.

"Hold it. What??"

Malachi closed his eyes and pursed his lips in irritation.

"You had promised not to interrupt."

"I know, and I'm sorry. But, seriously. What the actual fuck? I don't have these powers."

"They lie dormant in you."

"Then who's 'keeping the balance' or whatever?"

"There is an active Gatekeeper, but she is coming to the end of her time. Once she passes through the veil, you will come into your powers. And that," he continued, forestalling my next question, "will be any day now."

"Always two, there are?" I asked with a terrible Yoda impression.

Malachi looked at me strangely before letting it go and nodding his head.

"There are always multiple potential Keepers. Like prophets. Going about their days with no idea of the power they house."

"Then how do you know I'm next? Isn't it possible some other person in Boca Raton or something is on deck?"

"Boca Raton?"

"First far-flung place I could think of."

"Aha. No, it is you."

"But how do you know that?"

"Because you were created specifically for this time. To combat Damien and the apocalypse, he will start if left unchecked. His pairing with The Crone spells extremely dire tidings."

"The Crone?"

He nodded.

"An extraordinarily powerful and dangerous witch. She's been around long enough to be classed an 'ancient' now."

"Like Isaiah?"

"Much like Isaiah, yes. They were formed around the same time. In fact, he is currently working tirelessly to try to undo what The Crone did to imprison you."

"Oh. I see. I think." I shook my head. "This is a lot to absorb."

"Well, get over it quickly. Callisto and Damien have a bit of a head start on us. So, it's time for you to awaken."

He smiled for the first time since we met.

"Now."

I awoke and sat up with a heaving gasp, my forehead colliding solidly with Kaden's.

"Motherfucker that hurt… Oh! Motherfucker!!" Kaden pounced and threw his arms around me. "You're awake!"

"Hi," I responded, muffled by his chest.

Suddenly, Kaden was gone, and I was immersed in a scent I knew and loved better than anything.

"Oh, thank God, Gelli. I'm so glad you're back, baby," Draco whispered into my hair as he held me tight.

"How long was I out?"

"Close to three weeks." Draco held me tighter and kissed the top of my head. "Almost three weeks, honey."

"It's good to see you're back with us, Angelica," Isaiah said from against the far wall. "You had us worried. Especially considering The Crone is involved in all of this."

"That reminds me! I have so much to tell you guys… The Crone, Damien, Malachi -"

"Wait, Malachi is involved as well?" Isaiah scrubbed a hand over his face. "That's a possible bit of good news."

"Only possible?" Kaden asked.

"Possible," Isaiah nodded. "He is not exactly the most predictable of beings."

"Nice to know you remember me, Ancient," Malachi interjected, materializing into the room.

"Malachi," Isaiah greeted him, "A dubious pleasure."

"Indeed."

“Anyone else gets the feeling we’re missing giant chunks of this conversation?” Kaden asked, watching the exchange between the two timeless beings.

Half The Battle

Damien and Callisto lay panting on her kitchen floor.

"All is not lost, demon. Though the plan to trap her is compromised, you still have me in your camp."

"There is that."

Despite facing odds more difficult than he had earlier that day, Damien felt satisfied. Maybe it was the sex.

"The Cage did produce some interesting intel. She is still very much in love with the mage. We can use that," Callisto said, stroking Damien's wounded horn.

"Rache did say something to that effect."

"We could try something similar with him. Trap him in a spell."

Damien was shaking his head halfway through her sentence.

"No more hitting them from the shadows. I'm regaining my strength, and it's just a matter of time before I slip the veil and travel between planes. We are going to take this fight to them and destroy them."

"So, we have a plan."

Callisto giggled and grabbed his crotch.

"Let's celebrate."

~

"Try again."

Malachi had begun training me to create golems. I campaigned to get Isaiah to do it, but it was decided by consensus he was better used in the library. Yay me.

"I'm just not getting this, Malachi. Can't you do some angel-demon mojo and unlock my superpowers?"

"Superpowers?"

"Yeah, the Gatekeeper magic. You know," I waggled my fingers at him, "poof, demon-be-gone."

"Your irreverence is taxing."

My shoulders slumped.

"I'm just tired of failing at this."

"You're not failing," he said, setting his coffee cup down, "you're learning. Haven't you ever learned something before?"

"I have had a tendency to give up when I'm not automatically good at something." My mouth dropped open. "I've never said that aloud before. What the actual fuck?"

He had the audacity to laugh.

"That is one of the effects I have. Being around me for extended periods of time and it's like… What is that Amazon woman called? Her lasso of truth."

"Wonder Woman? Did you just compare yourself to not only a woman but a DC character on top of it?" Now, it was my turn to laugh.

"Anyhow," he continued as if I weren't laughing at his expense, "this is one time you do not have the luxury of giving up. You just have to keep learning until you get it." He shrugged. "Unless you're satisfied living in a world ruled by Damien and his Mate."

“You didn’t mention before they were mated. What does that even mean?”

He picked up his coffee and sipped contemplatively.

“You’ve heard of soulmates? Well, such things are very real. And like everything else, because a balance must be sought, there is an infernal version, too.”

My mouth dropped open for the second time that day.

“Are you saying he has a partner now?”

“Unfortunately. And for more good news, their being mated enhances each other’s power. It’s the reason you were created. To have a chance.”

“No pressure.” My stomach turned queasy, and a chill ran through my body. “I think I’m going to be sick.”

“We can take a break.”

He stood and muttered to himself as he walked back to the house.

“Such fragile creatures to pin the World’s hopes on…”

I awoke the next morning to every molecule of my body in pain. Between the training with Malachi and the sparring with the boys, my body was being put through the gauntlet. The bedclothes rustled behind me.

"Good morning, sweetheart," Draco mumbled, nuzzling the back of my neck.

"Mmmm… Good morning to you." I turned to face him, grunting with the effort. He started chuckling.

"Feeling the effects of training a bit?"

"Oh, shut it. I hurt too much to try to kick your ass right now."

"Speaking of a black eye first thing in the morning…"

"No! Don't remind me!" I started laughing, remembering the time he had tried to awaken me in an especially pleasant manner, and I awoke suddenly, swinging an elbow into his face. "You were so proud of that shiner."

"It was a love tap from my lady. What's not to be proud of?" He started tickling my ribs, and I tried to fight him off.

"Ow! No! It hurts too much to laugh right now!"

He stopped tickling and rolled me onto my back, resting himself between my thighs.

"Where does it hurt?"

“Hmmm… Everywhere.”

“Oh? Does it hurt here?” He punctuated his question with a thrust of his hips.

“Oh. Different ache starting there…”

“Is that so? Anything I can do to give you relief?”

“There may be,” I replied with a teasing smile.

He returned my smile and propped himself up on his elbows, caging my face with his hands. He ran a gentle thumb along my bottom lip before leaning down to kiss me. He broke the kiss and looked down at me, smoothing my hair away from my face.

“So many ideas of what I’d like to do to you, Angelica,” he murmured in a husky voice. His pupils dilated, and I felt his pulse against me.

“Is that so? Tell me.”

“I’d rather show you.”

“I have ways of making you talk.” I gave him a saucy look.

“Darling, I was the torturer-elect for a bloodthirsty coven,” he responded with a smirk.

“So?”

"So…" He lifted onto his hands and positioned himself lower.

"I have ways of making you scream."

~

"Let's try something different today," Malachi said as he stretched his neck.

"I'm up for a different tack," I responded, "I know you said I'm 'learning,' but I'm tired of feeling like I've shown to take a test I not only didn't study for, but didn't even realize was happening."

Malachi nodded and started walking away, indicating I should follow. He walked us over to a section of Isaiah's acreage with soft grass and shady trees. In the center of the clearing was a large circle of crumbling stone. The place felt ancient, and old magic tickled my skin.

"What is this?" I whispered reverently.

"The foundation of Isaiah's first dwelling. He built it on an intersection of ley lines long ago."

"Wow. It's amazing."

I studied the faded rune work and ran my hand across the rough face of the stones. Malachi stayed silent and watched me inspect the etchings. Energy ran the length of my body and hummed beneath my skin. I felt as if I were in the center of an electrical storm. He pushed off of the stone he was leaning against and approached me.

“Settle yourself in the middle of the stones and close your eyes.”

I obeyed and sat cross-legged in the center of the clearing. I closed my eyes, my hands resting on my knees, palms up.

“Now concentrate. Reach out with your mind and your magic. Try to find the ley lines beneath you.”

I fell back on meditation techniques, slowing my breathing and picturing my energy and magic connecting with the Earth. My heart rate slowed, and calm overtook me.

“No.”

Malachi’s voice startled me. I opened my eyes to look at him.

“This is not meditation. This is actively seeking an energy you should be able to find.”

“Oh, okay. I’m sorry.”

“Don’t be sorry, be better.”

I nodded and closed my eyes again.

"Try and see the magic surrounding you first. The circle of stones contains the energy. Let your magic reach out and interact with it."

I concentrated on the tickle along my body, which I felt when I entered the circle, and allowed the power within me to stretch and seek. Warmth spread in oranges and yellows from my core and leached through my skin. Small tendrils of blue and white crisscrossed lines wrapped themselves around the edges of my magic.

Focusing my attention downward, I sensed a pulsing power beneath me, beating like a heart. I allowed my essence to seep into the earth, seeking that river of silvery energy. I reached out with my magic and touched the river.

A massive jolt punched me in the chest and sent me sprawling in the grass. My eyes flew open as I landed flat on my back. Panting, I lay there, literally and figuratively shocked.

Malachi's face appeared above me, gazing at me upside down. He chuckled.

"I told you to find it, not touch it." He ran his hand over the stubble on his jaw. "But very good," he nodded. "Very good start."

I gave him a thumbs up, still too dazed to talk.

"Get up and try to move. Touching a ley line can be disorienting. See if you can walk."

I placed my hands beside me and pushed myself into a sitting position. My head swam as if I'd been drinking. I waited for the spinning to stop before I tried to stand. Gently, I rocked to my knees, placed one foot in the grass, then another, and slowly stood.

"There you go. Take it easy for now."

Malachi offered me a bottle of water. I sipped it, my stomach a bit queasy.

"Do you want me to try again?"

His face cracked into a smile. It was only the second time I'd ever seen him smile, and the effect was electrifying. One thing about both Damien and his brother is that they were exceptionally handsome in their human form. I didn't know if Malachi had another form, and I wasn't keen to find out.

"No. We'll try again tomorrow when you've had a chance to recover some. For now, go back to the house and eat something."

He dismissed me with a wave of his hand. I was happy to comply. Still slightly unsteady, I made my way back to the house where Kaden and Draco were probably waiting for lunch.

Draco met me on the front lawn. He pulled me into his arms and kissed me sweetly.

"How is training?" He brushed my hair away from my forehead.

"Better. I think. Hey, what do you know about ley lines?"

He smiled and placed his arm around my waist. I leaned into him, grateful for the support.

"I've done quite a bit of ley line work. My time with the Coven was spent training in the art. Is that what Malachi has you doing now?"

I nodded, and he squeezed my waist.

"That first time is a rush, right?"

"That's one way to put it. I guess I wasn't supposed to touch it. It launched me across the grass."

He huffed a laugh.

"Isaiah would settle near some powerful ley lines. Don't worry. It gets easier once you know what to expect."

"I hope so. Even though it punted me, it was the first time I felt a little hopeful I could do something to satisfy Malachi. He actually smiled today."

"No kidding? What was that like?"

"Dazzling." I smiled and pulled him down for a kiss. "Let's get some lunch. Now I'm feeling more steady, I realize I'm starving."

"Good, 'cause Kaden has been in the kitchen all afternoon, whipping something up."

"Truly? I thought Isaiah just had a magic kitchen which provides whatever he wants."

"It serves as a normal kitchen, as well." He kissed the top of my head, and we walked to the house, hand in hand.

The next morning, I woke up feeling amazing, like I had a battery charged. I felt like I could fight the sun. I nudged Draco.

"Draco… Drake?"

He mumbled something and pulled the blanket over his face. I looked at the clock. 4:47 a.m. Maybe a touch early. I popped out of bed and started to put on my workout clothes.

"Where are you going, sweetheart?" Draco's hair stuck up in six directions, and he peered up at me with one eye as he sat up.

"I feel incredible. I need to go for a run."

"Ley lines," he yawned.

"Whatever it is, I'm all for it."

"Okay, baby. I am not going with you."

"I understand completely. The sun isn't even up yet." I gave him a wide smile. "I love you."

He nodded and lay back down, snuggling deeper into the covers.

"Love you, too," came his muffled reply.

It was still dark as I vaulted off the front porch and onto the front lawn. Over the previous weeks, I had gotten a pretty decent lay of the land and started mentally mapping out my jogging route as I stretched. There was a river which cut diagonally across the property, and, heart light, I aimed for that.

The birds were still quiet, and the early morning chill was refreshing as I jogged. I started at an easy pace. Running had never been my passion, but I understood those who loved it. When I found a good rhythm, it felt graceful and heady.

I reached the river and turned to run along the bank. The first birds began waking up as the gray predawn lightened the sky. I picked up the pace, and a pleasant burn settled in my chest. I admired the flowers dotting the riverbank and smiled at a cute bridge crossing the river.

I ran for a few miles before deciding to turn around. The sun had crept over the horizon, illuminating a massive willow with long branches swaying in the current. I circled the tree and started to make my way back. My shoes whispered over the grass, and I had a decent sweat going.

On my way back, I caught sight of the ring of stones on my right. Naturally, I turned in that direction and trotted up to the clearing, stopping just before entering the circle. I stretched against one of the stones, eyeing the spot where I had sat the day before.

The sun warmed my back as I regarded the clearing. Apprehension warred with curiosity. Part of me cringed in fear of the power, while a larger part was eager to try again.

I examined the fear and realized I was equally fearful of success as failure. What if I couldn't do this? What if I could? The unknown stretched ahead of me like an unlit road.

"Thinking about trying again?"

I started, surprised by another voice in my peaceful morning. Malachi stood a few paces behind me, studying me with his arms casually crossed. I was so wrapped up in my thoughts I hadn't heard him approach.

"What if I fail?" I turned to face him, my back to the clearing.

"The world will burn, subjugated beneath my brother's heel."

"Again, no pressure."

He shrugged.

"What if I succeed? What would that look like?"

He regarded me for a moment before he spoke. "Then life will continue."

"But wouldn't it change everything? My being a 'Gatekeeper'? Won't my life be upended, and a new path need to be taken?"

"Only if you wish it so. Being a Gatekeeper is a responsibility, not a job. It's not as if you wake up, take a shower, and punch in. It is who you are. You were created to help keep the balance. The Apocalypse doesn't happen every time a cheeky little demon breaches the Veil. This time is different."

"Because it's Damien."

"Yes."

I nodded in understanding. With that explanation, my fear was almost fully supplanted by a budding pride. For a moment, I basked in Malachi implying I was special.

"But for now, you are a novice, beginning to learn what being a Gatekeeper means and what is expected of you. You are not even close yet. And time is not on our side."

...And we were back. He gestured toward the circle.

"Are you ready to try again?"

"Yes," I said without hesitation.

"Good."

I turned back toward the stones and strode into the center, crossing my legs as I sat. I closed my eyes and reached out with my magic to find the blue and white energy buzzing all around me. Entwining the two, I then focused on the earth, seeking the silvery river I had seen yesterday. Hesitantly, I allowed the orange and red of my magic to approach. It hovered a hair's breadth from contact.

This time, instead of plunging into the river, I let the river reach out to me. It felt almost curious. Tendrils of pure energy crept toward my magic and flirted with contact.

When it wrapped around my proffered power, it wasn't a punch to the chest this time. It felt like holding a live wire, just shy of uncomfortable. My body lit up from the inside, causing me to raise my face and breathe in shallow gasps.

"Breathe through it. That's it. Keep breathing."

Malachi's voice came to me as if through water, muffled but present. I tried to take slow breaths despite my pounding heart.

"Okay, now break contact."

I slowly pulled my magic back, and the river released me. My eyes popped open when my butt hit the ground. I looked over to Malachi, who appeared pleased.

"What the hell? Was I floating?"

"Just a little."

"Well…*that's* awesome. Does that mean if I get the hang of ley line work, I'll be able to fly?"

"I suppose you could. It's not common, as you can imagine. And it's not the purpose of our work right now. Right now, I'm teaching you how to pierce the Veil and push beings through without allowing others to escape. It's delicate work you need to master for it to be useful."

"I'll do my best."

"You'll have to do at least that."

Malachi accompanied me on my walk back to the house. We spoke little, each wrapped in our own thoughts. When we arrived, we found Kaden, Isaiah, and Draco in the kitchen, having breakfast. Even Mischief was there, snacking on some premium cat food.

"How was your run?" Draco asked around a mouthful of eggs.

"Invigorating. I also did some ley line work. I feel I might be getting the hang of it."

"That's excellent news, young one," Isaiah remarked, giving me a fond look.

Ever since my spiritual coma, Isaiah had shown a marked difference in his regard for us. My heart filled with gratitude at the thought of him in our corner. Hope bumped against my heart. I wasn't ready to believe we could defeat Damien just yet, but the weight of the responsibility was now shared with some big players.

Famished after my run and the energy exchange with the ley line, I grabbed a plate and began loading up.

Why were you running? Was something chasing you? Mischief's voice echoed in my mind.

"No, sweet girl. I went for exercise."

Oh. Like when I chase feathers or string, Mischief nodded and resumed munching.

Kaden chuckled. The energy in the room felt companionable and safe. For the first time in what seemed like years, I felt okay.

~

Damien seethed with frustration. His brother - the eternal thorn in his side. And that insufferable witch and her pompous family, with the two mages at her side. Without a solid plan, doubt was creeping in.

His ability to jump through planes was almost back to full power. Now, it was a waiting game. Damien hated waiting.

Even the fantasies of destroying her were losing their novelty. He wanted the real thing. Itched to hurt her in every way possible. He was so close. Every day he was closer. But not quickly enough to satisfy him. He wouldn't be satisfied until he had her flesh in his teeth.

Calliope and Siren were in her kitchen, putting spells together for clients. She occasionally allowed people to purchase her wares, as long as she approved of them and their intent. Sowing discontent and chaos.

Siren slunk into the living room like a phantom. Damien glared at her, and she growled back. They would never be friends, but they had reached a detente.

"Are you two getting along?" Calliope called from the other room, laughing at her own joke.

She found endless delight in their antagonistic relationship. She made it clear whose side she was on. When it came to Siren, Damien could pound sand. Infernal mating was about power, not love.

"Like a house on fire," he responded, showing the cat his middle finger—a distinctly human expression, but one he found funny.

Damien was bored. Without the one thing he wanted, everything paled in contrast. Food tasted like sand, and sex felt perfunctory. Even torturing and consuming souls felt like going through the motions.

"I'm going out," he called toward the kitchen. Maybe a visit to a tabernacle would cheer him up.

He rose from the chair and sneered at Siren when she growled at him. He was halfway out the door when Calliope called his name.

"What? I had one foot out the door."

"I have something which may please you."

He grunted. "Stop torturing me, woman," he barked. He was in no mood for one of her many, many games. A chill breeze surrounded him.

"You will watch your tone with me, demon," she said, her voice edged in ice. "Besides," she continued almost cheerfully, "this is something we will both enjoy."

She came around the corner, wiping her hands on a towel, and then whipped it over her shoulder. He regarded her expectantly. She smiled slowly, satisfaction glinting in her eyes.

"Well? Am I to guess?" His voice took a dangerous edge.

"I think I may know where they are."

Damien strode quickly to her until they were nose to nose. He grabbed her shoulders and squeezed tight. "If this is another of your ridiculous games…"

She giggled, her pupils dilating.

"Mmm. No game."

He released her and began pacing the parlor. She shrugged and sat in her rocking chair, her hand resting on Siren's neck.

"How could you know where they are?"

"Logic and a long life."

He continued pacing but shot her a dark look. She rocked casually, then rolled her eyes.

"No normal concealment ward is strong enough to fool me. We are not able to track them, which suggests they are being cloaked by a ward undetectable even to me. That would suggest an Ancient." She stroked Siren as she spoke. "There are only a small handful of those left."

Damien stopped pacing in front of her and bent down to kiss her hard. "That's the greatest thing you've ever said. You are my muse."

"Show me."

Damien growled, grabbing her and throwing her over his shoulder, surreptitiously pushing Siren away with his foot.

~

To my relief, Malachi had me focusing solely on ley line work, leaving the golems to Isaiah for later. Since The Crone got involved, Isaiah seemed fully on board. The only silver lining to Damien having a partner.

I was sitting in the center of the stones, feeling the energy course through me, when Malachi appeared, carrying a burlap sack. The sack was moving; something struggled inside. He tossed it to the ground unceremoniously. The sack grunted. I eyed it suspiciously.

"Um, what's going on, Malachi?"

"New stage of your training. You're going to try and send a demon back through the Veil."

Anxiety flooded me.

"What? No way I'm ready for that."

"You won't know until you try. It's just an imp. Easily manipulated, and I'll send him through if you can't."

"I thought I was the only one who could do that."

"You are a Gatekeeper. You're the only one who can send demons like Damien through. Small beasts can be handled by any angelic or infernal being."

"I don't know…"

"The current Gatekeeper is reaching her end, Angelica. The time is now, whether you're ready or not."

I slumped, knowing my time to take up the mantle was fast approaching. I had the luxury of a little denial until that moment.

"Okay. What do I need to do?"

"Follow your intuition, mostly. You've been doing well with your ley line work. This is just another step in energy manipulation."

He hefted the bag into the circle of stones.

"We'll do this here because the energy is so concentrated. Once you get the feel for it, you'll be able to pierce the Veil anywhere. It's all around you. Some gateways are easier to see than others, like near ley lines. But as a Gatekeeper, you'll be able to open one anywhere."

"That's good. I was wondering about that. I doubt Damien would be considerate enough to battle us near a strong ley line and make this easier. Okay," I said again, "what's my first step?"

"For now, close your eyes and tap into the ley line. Once you've connected, open your eyes."

In the weeks I'd been practicing, I'd never opened my eyes during energy work. I assumed I wouldn't see the energies without my eyes closed. The irony made me chuckle internally.

I straightened up, legs crossed, and closed my eyes. I found the silvery river with no problem and linked my magic to it. Then, I opened my eyes.

"Holy wow…" I breathed.

The clearing was a network of crisscrossing lines and pockets of energy, shimmering with power. It was stunning. I turned, looking all around me, noticing some areas shimmered more intensely than others.

“Do you see it?”

“I see a lot of lines and pockets of concentrated power.”

"Good. Those 'pockets,' as you say, are gateways. You'll notice some are stronger than others, yes?"

“Yes."

“Those which burn brighter are larger gateways. You can use any, including smaller, weaker gateways for larger demons, but it is a lot more difficult to get them through." He opened the sack. The imp stumbled out and immediately started running.

"Stop him with a rictus spell."

I obeyed, accidentally breaking my connection to the ley line and the lines and pockets disappeared.

"Damn it."

"That’s fine. He’s not going anywhere."

I smothered my frustration and tried to connect again. It felt like rubbing my stomach while patting my head. One hand was holding the creature in a rictus spell while I reached out to the ley line with my other.

I connected, and the field lit up in a miasma of color and light again.

"Very good. Use one of the larger ones for the purpose of this exercise."

"How do I get one to open?"

"Are you connected to the ley line?"

"Yes."

"Good. Treat your body as a conduit and send the energy toward one of the bigger gateways. With sufficient force, you'll be able to split it open."

Out of habit, I closed my eyes to concentrate.

"Keep your eyes open. You need to see where you're sending the ley line energy or it could go anywhere. It has the potential to be very destructive."

"Right."

I focused on the biggest pocket I could see and concentrated on bringing the ley line energy through my body. The sensation was tremendous, like swallowing lightning, making it hard to breathe.

"Keep breathing slow, deep breaths," Malachi instructed.

I nodded, struggling to draw in a deep breath while balancing the effort of holding the imp in the rictus spell.

Sweat beaded on my forehead, and my arms shook with exertion. After a minute or two, I started to get the hang of it, pulling energy from the river into my core. One hand remained connected to the ley line, and the other gripped the imp.

"Um, Malachi?"

"Yes?"

"What do I do now? I've run out of hands."

"Pivot the hand you're using to connect. If you've drawn enough in, you should have plenty to open the gateway."

"Okay."

With hesitation, I broke my connection with the ley line. As Malachi predicted, I still felt electrified from within. Turning my arm, I aimed toward the pocket and pushed. A bolt of silver magic shot from my hand straight toward the target. Initially, the energy splashed against it, and I feared failure. Then, the middle of the stream began to split the pocket open, tendrils of inky darkness creeping around the edges.

"Send him through. Now!"

I whipped my other hand around and aimed the imp toward the split in the Veil. With a squeak, he disappeared, and the seam sealed itself. Once again, I was looking at a grassy clearing, sunlight dappling the leaves on the trees around us.

“I did it!”

“You did.”

“I can’t believe I - oh… Uh oh…” I slumped forward, suddenly boneless and bereft of energy. Malachi stepped forward to keep me from face-planting into the dirt.

“Very solid work, Gatekeeper.”

“Why do I feel like I’ll never have energy again?”

The colors around me seemed muted, and the sun appeared weak.

“That is what some practitioners refer to as the ‘hangover.’”

"Oh. It sucks."

"That's what I understand."

"How can I fight like this? I'll be helpless and vulnerable."

"You used a massive amount of energy to pierce the Veil. In time, you'll be able to dole it out in more precise amounts, keeping some for yourself. For now…"

He lifted me and cradled me in his arms. "I'll just carry you back to the house."

"My hero."

He rolled his eyes.

The next day, Malachi had me sparring while trying to connect to a ley line. The boys were taking it easy on me, but I'd still been ending up on my ass all day. Splitting my focus between fighting and establishing an unbroken connection with the river of energy had my head spinning. Draco held his hand out to help me up from the ground again.

"Maybe we can take a little break," Draco said, giving Malachi a hopeful look. Malachi sighed.

"Yes, that's fine. 15 minutes."

I gave Draco a grateful smile as he pulled me up from the grass. Kaden handed me a bottle of water.

"Thank you, boys. What would I do without you?"

"You'd spend less time on the ground and be thirsty."

"Ha ha."

We rested and stretched for a bit before Malachi called us back to the clearing.

"We'll try this for a few more minutes before calling it a day."

"Sounds good, coach," Kaden said with a sharp salute. I rolled my eyes at him.

Kaden and I squared off. I reached for the ley line and connected. Kaden threw a runed spell at me, and for the first time, I blocked it without losing my connection.

"I did it!" I shouted, elated.

"Good job," Malachi called from his spot against one of the stones, "Now try to send this guy back."

I set my feet shoulder-width apart, blocked another spell, and grabbed the imp running from the clearing. Breaking my connection, I shot a spell at Kaden, then the gateway, hitting Kaden in the chest and splitting the Veil. I tossed the creature through, and the seam closed.

I felt as if I could do backflips. I had never learned how, but I felt like I could. Draco ran to me, picked me up by the waist, and spun me around. Kaden jogged over, rubbing his chest.

“Nice work, Pepper,” Kaden said, clapping a hand on my shoulder.

“Good job, baby,” Draco said, kissing me hard.

“Thank you! I finally did it!”

“Yes, yes. Good work,” Malachi added with a ghost of a smile, “now you need to be able to do that every time. But it’s a good start.”

More than anything, I felt relief. I had begun to think I’d never get the rhythm of fighting and ley line work together.

“Hey, Malachi?”

“Yes?”

“You once mentioned ley lines could be used for destructive purposes. Do you think I should start learning how to use them offensively? Would that be helpful in a fight?”

"Maybe. We'll talk about it tomorrow," he said, before turning and walking away.

Once he was out of sight, I bounced on my feet and clapped my hands like a little girl.

"Really great job, baby. I'm so proud of you." Draco threw an arm around my shoulders and kissed the top of my head.

"Thank you. I was seriously beginning to think I'd never get it. This gives me some hope I'm not totally lame."

"Well, you're not totally lame," Kaden said with a smile.

Suddenly, I was overcome with fatigue and hunger. I let out a breath and asked, "Hey, can we go back to the house? I'm starving and feel like I could sleep for three days."

"You got it, love." Draco gave my shoulders a squeeze before dropping his arm to take my hand.

When we got back to the house, we found Mischief chasing a string Isaiah was dragging along the floor. She pounced.

Got it!

She picked up the end in her mouth and began to walk away, head held high. Isaiah dropped his end and let her have it.

"She earned it," he mouthed to us before covering his mouth with his hand, stifling a giggle.

Isaiah had done a complete 180 since the first day we met. He was now lighthearted and kind, not the curt Ancient who had slammed Draco against the kitchen wall the first night we were there. It almost felt as if he were enjoying having company. Eons alone, just "collecting" information about the world without engaging, must get lonely. He was even playing with Mischief. The thought had me smiling into Draco's shoulder.

Release The Hounds

The Crone had set up six locator spells for the remaining six Ancients: Esme, Isaiah, Micah, Lois, Mohammad, and Xing. She hummed while she worked. They had all crossed paths with her throughout the years. There used to be a dozen. She herself was supposed to have been the thirteenth.

Though not technically an Ancient, Calliope had been around as long as they had. She had been created to be an Ancient, but she never passed an Anointing. The Ancients were about love, learning, and magic. She was made of sterner stuff, just pure magic without conscience.

It used to stick in her craw that she was never accepted into their little club. As time wore on, though, she not only came to terms with it, she reveled in the fact that she was not one of them. If they were so great, how had she destroyed six of them? Power was the only thing which mattered.

"How's it going, kitten?" Damien leaned back and slung his leg over the arm of his chair.

"Don't hover. I'll let you know."

Even Calliope barking at him didn't wipe the smile from his face. He was in too fine a mood, getting closer and closer still. He could practically taste the witch's blood on his lips.

"Copy that. Going out."

She didn't look up from her work as he brought his foot back to the floor and stood up. From her place at Calliope's feet, Siren watched him from the corner of her eye.

"Yes, I hate the kitty," he whispered in baby talk, "yes I do." Siren lifted a lip to show him her canines and growled softly.

"Get out," The Crone said perfunctorily, without sparing a glance in his direction. "You're distracting me."

Damien saluted the back of her head facetiously and turned on his heel, striding through the door.

In a fine mood indeed.

The day was oppressive, muggy, and overcast. Damien turned his face skyward, noting rain was likely. He squished through the marsh, plucking bugs out of the air. He missed the cool, dry air of the northern regions.

The Crone's choice to live in this cesspool baffled Damien. She could reside anywhere, yet she chose a swamp infested with bugs and dense vegetation. He didn't care enough to inquire about her reasons, but he wondered on occasion. This climate too closely reminded him of Downstairs, a place he was eager to forget.

He never wanted to return to his birthplace. Given a chance to bring the world to its knees, he had failed once; he could not fail again. Especially with the Infernal Mating with The Crone. He was being armed with every weapon to complete this task, and he was loath to disappoint Legion again.

He shuddered at the thought of what his future held if he didn't succeed in bringing this world to heel. The only obstacles in his path were a young witch, her two mage friends, and one Ancient. He sneered.

He would not fail.

~

Isaiah had sequestered himself for the last couple of months, researching day after day and well into the night, looking for the right spell work to use against Damien. With The Crone involved, he wanted to be sure our battle tactics were foolproof. Isaiah held a healthy fear of her, and with Damien and Isaiah in the mix, he wasn't pulling any punches.

Once again, I felt deep gratitude he and Malachi were in our camp. Without Malachi and Isaiah, Kaden, Draco and I would be rudderless and terrified. I would never have figured out anything regarding being a Gatekeeper. I may have discovered new powers, but I wouldn't have known how to wield them or their purpose. As it stood, I was tirelessly practicing ley line work. Knowing better than to underestimate Damien, and seeing Isaiah's concern over The Crone had Isaiah scared, count me scared, too.

Alone in the clearing, I was practicing opening gateways when Malachi approached.

"Isaiah has something to share with us regarding battle spells," Malachi said by way of greeting.

I pushed a lock of white hair out of my eyes and stood, brushing grass from my skirt.

"That's great news. I mean, I never doubted he would come up with something, but he's been working on this for some time."

"Battling a pair such as those two, years of preparation might not be enough," Malachi replied gravely.

A spike of anxiety shot through me.

"That's a sobering thought," I admitted.

Malachi gave me a grim nod.

"It was very fortunate you chose to seek out Isaiah. He's unmatched in research and spell work. Or so I'm told."

"You guys have met before."

"Yes, a very long time ago."

A small frown appeared briefly on his face, immediately replaced by his usual mask of bemused indifference. He was not forthcoming with anything more, and I didn't ask. But there was definitely a story there.

We walked into the house, following the voices coming from the library. The overall tone was mostly lighthearted like friends catching up over lunch. Three sets of eyes were on us the moment we walked in.

"Oh, good. You're here."

Isaiah clapped his hands together and sat at a desk overflowing with books and parchment. I sat next to Draco on a loveseat. Malachi remained standing, arms crossed.

"I'll get right to it. The two we are up against are extraordinarily powerful. The Crone, or 'Calliope,' as she used to be known—"

Isaiah waved his hand sweepingly, and the image of a devastatingly gorgeous, dark-haired woman with bright green eyes appeared before us. Beside her, a perfect embodiment of power stood a sleek, black panther. The image morphed into a wizened old woman with stringy hair, dressed in rags and holding a staff. The panther still stood by her.

"—was originally created as an ancient. She can change between the three feminine forms: The Maiden, The Mother, and The Crone. She spends most of her time in the first and third forms, as motherhood never suited her. She is," Isaiah scrubbed a hand over his face, "formidable. The cat, Siren, is her familiar or her avatar, as it suits her. To incapacitate her for good, both forms must be contained or destroyed."

The image changed again to depict thirteen stunningly beautiful people.

“This is all thirteen of the Ancients, including Calliope, before she struck out on her own. Six remain. That one is me.”

He pointed to a strapping vision of manliness with coiffed hair in a three-piece suit. He was hot as all hell. Isaiah ignored the questioning stares and moved on.

“Over the millennia, The Crone has destroyed six of my brothers and sisters. As I said, she was originally created to be an Ancient, but it requires a level of moral fortitude she did not possess. What happened to her to result in her turpitude has been hotly debated over the years. No one knows why she never passed the Anointing ceremony—a ritual during which one is awarded the title ‘Ancient’—but she soon descended into the black arts, never to resurface. What is important to note here is she was created to be one of the most powerful beings in creation and should never be underestimated.”

Isaiah waved his arm again, and an image of a ramshackle house in the middle of a swamp appeared.

"This is where she resides now, deep in an American swamp. She's lived in many places over the eons but has been here the longest. Occasionally, she sells her services to those debased enough or desperate enough to seek her out. She is very accomplished in her craft. In fact, I'm mildly surprised she hasn't knocked on my door yet."

A soft, collective gasp escaped from Draco, Kaden, and me. The thought that Isaiah was at all vulnerable had never occurred to any of us. He waved a hand, and an image of Damien replaced the house. Goosebumps pimpled my arms, and Draco took my hand.

"You are familiar with this subject. He is a top demon in the Hierarchy and has been plaguing the Earth for some time now. He's been captured once, but without being sent back to Hell, it was always a temporary solution."

"Would've been nice to know that," Kaden grumbled from his spot on the couch.

The image changed to Damien's true form. A chill ran down my spine, and unease settled in my stomach.

"He can jump through the planes and appear or disappear at will. Exceptionally strong and... mean."

Malachi chuckled softly.

"Mean…" he muttered under his breath.

Isaiah scratched his chin. "Perhaps he would be better described by his brother."

Malachi shook his head from his post near the door.

"No, 'mean' is a succinct way to describe him. He revels in torture, loves virgins, the whole disgusting thing. He is mean. And dangerous. And loves every moment of it."

Isaiah waved for him to continue. Malachi sighed and dropped his arms. "He takes direction only from Legion, one of the founders of Hell. Refers to him as 'Father,' which speaks to his mental state. He is unhinged. But very good at what he does—raping, torturing, and killing—so the Court gave him an opportunity to try to subjugate Earth. It is the great obsession of every royal 'Downstairs,' as they call it, to overtake Earth. The weather is better up here."

Now it was Malachi's turn to gesture to Isaiah. Isaiah swept his arm, and an image of Malachi popped up.

"Hey! We know him!" Kaden piped up, pointing.

"Shut up, K."

"Aye, Pepper." He made a motion of zipping his lips.

“Malachi was created as the balance of Damien. And just as Damien has an infernal form—”

The image of Malachi in his true form was breathtaking. To call him beautiful was a disservice. He was perfection, from the tips of his toes to the tips of his wings. Massive, sleek, and luminous silver wings swept back from his sculpted shoulders. His hair was glossy black and tumbled past his elbows. His eyes were a myriad of colors, all heartbreaking. Draco reached over and snapped my jaw shut.

“-so Malachi has an angelic form.”

Kaden, Draco, and I all slowly turned to look at Malachi. He shrugged and crossed his arms again.

“So, you’re like, an actual angel,” Kaden breathed, his eyes wide.

“I’m a soldier,” Malachi answered.

“Of God,” Kaden added.

Malachi sighed. “Yes, of the Heavens. I was created to protect and fight for humanity. Damien was built to destroy it. Let’s focus, please.”

Isaiah hid a smile and turned back to the image of Malachi in his angelic form. After a moment, he waved his arm, and the image disappeared.

"I believe they will do one of two things. They will either track us here or try to draw us out. If they track us here, we have an advantage because of the proximity to the ley lines and knowing the land. If they try to draw us out, it will be to get us on more neutral ground. Frankly," Isaiah added, "the first option is, without question, the favorable one. Besides the advantage of 'home court,' what they would do to draw us out will result in a lot of bloodshed and mayhem."

Isaiah stood and began pacing.

"This is not Malachi's nor my first encounter with Calliope. We were there when she slaughtered Agatha."

I had to ask.

"Who is Agatha?"

Isaiah stopped pacing and exchanged a look with Malachi, as if asking for permission. Malachi shut his eyes briefly before nodding.

"Agatha," Isaiah began, bringing up the image of the thirteen Ancients again, "was a lovely creature and a very dear friend of mine."

He pointed to a willowy blonde with smiling green eyes. She stood gracefully beside Isaiah in the vision.

"She and Calliope were very close before Calliope turned her back on the Ancients. When she dove into dark magicks, Agatha volunteered to try and talk Calliope back into the fold. But Calliope was angry."

When Isaiah paused, Malachi uncrossed his arms and moved to sit in an armchair next to the fireplace.

"Agatha and I were in love," he said.

I tried to keep my expression neutral, the energy in the room suddenly became very heavy. Draco squeezed my hand, and I exchanged a glance with Kaden. None of us had seen that coming.

"I begged her not to go. I tried to warn her that Calliope had turned dark beyond help, but Agatha was so sure her friend was still in there." He shook his head. "The most she would agree to was allowing me and Isaiah to accompany her. She tracked Calliope to a villa in Marseilles and just walked up to the front door. Calliope opened the door and slaughtered her on the spot."

Malachi was silent for some time. No one moved. I barely dared to breathe.

"We had no time to process what had just happened when she turned her attention to us," Isaiah continued. "We had to act immediately, lest we be next. Malachi opened a portal and got us out of there quickly. We couldn't even collect what remained of our Agatha."

He scrubbed a hand over his face.

"That was a long time ago and early in her journey into the dark arts. We can only assume she's become even more ruthless and powerful. So, you understand why I've chosen to help you and why I take an attack from her so personally."

I nodded, loosening my vice grip on Draco's hand.

"So, it's a miracle she didn't kill me with that spell, then?"

"No. That spell was never intended to kill you. I think Damien sought her out to trap you and torture you at his leisure," Isaiah answered.

"If she'd wanted to kill you, she would have."

"So why doesn't she do it now? Wouldn't that end all of this?"

"I suspect Damien wants to be the one to do it. She doesn't have a personal stake in whether you live or die, so she does nothing. I think her spell on you was part of their courtship. She was showing off."

He sat down heavily in his chair again.

"Well. That's just peachy," Kaden said, standing and pacing slowly. "We have a top-level demon and a psychopathic witch bent on our destruction, and Pepper is only alive because Damien wants to kill her in person." He stopped pacing and faced Isaiah. "Do I have that right?"

Isaiah nodded.

"The good news is, I've put together some spells which may give us a fighting chance against those two. They are relatively straightforward but involve all of us working together. I believe it is quite doable with practice."

"Well, what are we waiting for? Let's get to work," Draco said, standing up.

Isaiah waved a hand. "It's already late, and we've given you a lot to process," Isaiah reasoned. "We'll start first thing tomorrow."

"That's fair," Draco responded, offering me his hand to help me up from the loveseat.

I stood and snaked my arms around his waist. He felt solid and warm.

The next morning, we all assembled in the clearing. Dawn was just beginning to creep into the sky, and the grass was wet with dew. I zipped up my hoodie against the morning chill.

"We have an advantage," Isaiah began. "We have each other. We can play to each other's strengths and mitigate the attack as a whole. Damien and The Crone are fiercely independent and used to working alone. Despite their infernal mating, I am reasonably certain they still behave as if they are autonomous."

We trained with a focus bordering on obsession. Isaiah was right when he said the spells were straightforward. All we needed was to perfect the choreography.

~

The Crone had completed the spells to locate the Ancients. Two were in Europe. Two were in the States, one was in the Middle East, and one was in China. She assumed the silly little witch and her friends weren't traveling outside the U.S., so she ruled out the overseas locations. She focused on the two in the States.

That left Isaiah and Esme. Easy. Next, she cast a spell to begin dissolving their wards. She anticipated they would be difficult and relished the challenge. Since Damien had come to her, she had been able to stretch her magic. It had been a long time since she'd had this much fun.

"You're smiling," Damien observed. "I like it when you smile. That means you've finished something."

"I've narrowed their location to two possibilities," she replied.

Damien grinned and grabbed her around the waist. "That's my little witch!"

The lights in her house flickered.

"You dare condescend to me? I should kill you," she hissed.

Damien lifted her and spun her around. Siren rose and began slowly circling them.

“Oh, stand down,” he said, rolling his eyes at the cat. “This is cause for celebration.”

He kissed her hard. She bit him hard enough to draw blood. He lifted her and wrapped her legs around his waist so she could feel his cock straining against the seam of his pants. She purred in his ear and licked his neck.

Siren went outside.

Later, The Crone busied herself crafting spells to interfere with the wards set up by the Ancients. She knew it would be tough but was confident she’d break through at least the first couple of barriers by the end of the week.

~

Today, Isaiah introduced the golems, our first regiment and primary line of defense. Made of earth, each one was huge and wore the same vacant expression. They were unsettling.

“As much as you can, use the golems to intercept anything thrown your way,” Isaiah instructed as he brought another creature to life.

Watching an Ancient craft was mesmerizing. His movements were so fluid, it was almost like he was dancing.

He crafted five golems, one for each of us to practice using as a shield. We threw spells at each other, bobbing and weaving. If it weren't so deadly serious, it would have been fun. But the fight against Damien and The Crone would be anything but. A pall of dread followed me everywhere these days.

A massive crack fractured the air, prompting me to duck for cover. Isaiah motioned for us to stop and tilted his head, his brow creased.

“The Crone has breached my first protective ward,” he announced gravely.

I felt as if the ground had fallen away. "Does that mean she knows where we are?" I asked, my voice trembling.

“Okay, baby, it's not time to panic yet.” Draco wrapped his arms around me, comforting me.

"We're not ready to take them on. "

“Not yet,” Draco reassured, rubbing my arms, “but we will be. Remember, we have some powerful folks on our team, too.”

I nodded, burying my face in his chest for reassurance.

Isaiah swept his robes behind him and strode toward the house. Kaden, Draco, and I exchanged worried looks.

"That's bad news," Kaden muttered, shaking his head slowly.

"Given a couple of weeks, she may penetrate all of his defenses," Malachi added grimly.

"Does that mean she knows where we are?" I pressed anxiously.

"It's a distinct possibility," Malachi confirmed, then turned to follow Isaiah.

I extricated myself from Draco's arms and marched towards the clearing.

"What are you doing, baby?" Draco called after me.

"If Damien and The Crone are on their way, I want to be damn sure I can send his ass to Hell," I declared firmly.

"You want company?" Kaden asked, determination in his voice.

"Absolutely."

While Malachi and Isaiah worked on repairing the fallen ward, I dedicated the day to practicing and harnessing ley line energy until Kaden and Draco had to practically drag me to get something to eat. I could now comfortably hold and distribute the magic like a faucet. I could blast a giant stream or send the tiniest of whisps. I could split the Veil into a window or an airport hangar with ease—all without collapsing with a magic hangover.

Manipulating magic was hungry work, and I ate my roast beef and tomato sandwich on rye with gusto. After today's effort and some food in my stomach, I was feeling satisfied. Later, I was so exhausted, I fell asleep in the library with a book on ley lines still in my hand. I woke when it fell to the floor with a thump and found Draco sitting in a chair next to me and watching me with a smile.

"Good morning, sunshine," he greeted with a wink.

"Morning," I mumbled. "Wait, is it morning?"

I sat up and scanned the library for a clock. It was windowless, which protected the books and gave the sense of existing out of time. It was one of the things I loved about Isaiah's library. Any time I wanted a respite, all I had to do was start poking around the endless rows of beautiful tomes.

"It's 11:13 p.m., sweetheart," Draco informed me.

“Oh,” I sighed, relieved I hadn’t slept through until tomorrow.

Draco closed the book he was reading on Runework and placed it on a side table. Standing, he offered me his hand. I took it, and he pulled me close. He smelled amazing, like rain and fresh earth, and I sighed contentedly into his chest.

Hi, guys.

“Mischief! Where have you been? I feel like I haven’t seen you in forever.”

I reached down to pet her, and Mischief flopped over. She righted herself and leaned into my hand.

I’ve been exploring the grounds. There is so much to find around here. No mice, though.

“Well, lucky for you, there’s a magical kitchen that can produce anything you want,” Draco remarked with a chuckle.

No In ‘N’ Out “wild style.”

"Not sure Isaiah has been out to have a cheeseburger in a very long time, sweetie," I told her.

Probably not. Are you learning, mama? Is that why you're in the library?

"I am learning. Would you like me to find you a book on healing?"

Could you? The ladders make me nervous, and there are so many books.

"I would love to."

Mischief bumped my hand again before going very still, eyes wide. Another ear-splitting crack shook the air, and Mischief jumped before shooting under the couch.

"Oh, crap. That's another ward, isn't it?" I asked, already knowing the answer.

Draco nodded and pulled me into another hug.

"Time is running out." He held me closer and pressed his lips to the top of my head.

Isaiah had locked himself into his library again, this time in an effort to counter the attack on his wards. A breach seemed inevitable. All he could do was buy us more time. We were all gathered in the study outside his library when he emerged, a prehistoric cordless phone cradled between his ear and shoulder.

"...Okay… Yes. You, too. 'Bye."

He pressed the button to end the call and faced our nonplussed expressions.

"What?"

"You have a phone," Kaden stated, pointing at the relic in Isaiah's hand.

"Well, I can't read telegraph."

"Yes, you can," Malachi said with a faint smile.

"Fine, yes, I have a landline. Not what I want to discuss."

He swept into a chair in the study and crossed his legs.

"That was Esme. Her wards have been attacked, as well."

"Interesting," Malachi muttered, rubbing his chin.

"I was thinking… Could you do a boomerang spell on the wards? Send the attack back to The Crone?" Kaden asked, sitting in a chair next to Isaiah.

"That would be ill-advised. It's like firing a shot in a combat situation. We would give away our position. The good news about Esme also being attacked is The Crone has not focused all of her attentions on my place, meaning she is not sure whether you are here."

"If Esme is being attacked, does that mean she may want to join us?" I asked, hopefully. Having one Ancient was incredible; having two would be invaluable.

"Not likely. She's even more protective and isolated than I."

"But it's a possibility?" I pressed.

"I suppose, but we can't plan as if she will."

"That's fair," Draco said, taking my hand. "We'll just continue as we have been."

Isaiah nodded.

The morning was crisp and cloudless. The sun warmed my skin as I sat in the grass, watching Kaden and Draco spar. After spending the last three hours working with the ley lines and gateways, I felt I deserved a small breather. Kaden and Draco also chose to take a break and joined me.

"How are you feeling, Gelli? Are you feeling better about the ley line work?" Draco asked.

"Much better. Well, you've seen a lot of my progress. Given enough time, I think I'll be able to fly."

"You'll be able to *fly*??" Kaden asked, his tone tinged with obvious jealousy. "Can anyone working with ley lines do that?"

"It seems to be something unique to me, I guess. Malachi said the other Gatekeeper can't do it."

"That is so excellent." Kaden leaned back on his hands and turned his face toward the sun.

"Why don't you try it now?" Draco asked, picking a blade of grass from my sweater.

"Yeah!" Kaden sat up childlike enthusiasm bubbling over. "I wanna see!" I laughed.

"Okay, okay. I'll try now."

I stood, brushed the grass from my pants, and made my way to the ring of stones. Settling in the center, I closed my eyes. The silvery river of energy rushed to meet me, making me feel as if I'd eaten a bowl of lightning. I manipulated the energy to gather beneath me and concentrated on it, lifting me off the ground. I imagined the whorls of magic swirling under me and supporting me.

"Holy wow!" Kaden exclaimed.

I opened my eyes at his voice and looked down. I was a good six feet off the ground.

“Can you move, too?” Kaden was running circles around me like a puppy.

“I don’t know, I’ve never tried. I’ve been concentrating on perfecting what I need to fight Damien.”

I tried to imagine the energy pushing me to one side or the other. Nothing happened. Eyes open, I watched the river of energy running beneath me and tried to bring it up to touch my waist and push my right side. I scooted two inches to the left.

“You did it!” Now, both Draco and Kaden were running in circles below me.

“Keep going!” Draco shouted up at me.

I pictured the magic holding me like a giant silver hand, carrying me to one side or the other. I swung to the side, listed, and fell. I hit the ground with a thump, knocking the wind from my lungs.

“Oh, man. Are you okay?” Kaden ran over to help me up. “That was so cool!”

“I’m fine,” I gasped, then grinned. “I flew!”

“Yeah, you did,” Draco said, grabbing me around the waist. He lifted and spun me in the air, kissing me on the way down. I felt incredible.

~

"The ward-breaking is working," Calliope called to Damien from the kitchen.

"Exceptional work, my little witch."

The lights flickered, and the temperature in the room dropped a few degrees. He chuckled. He knew how much that moniker pissed her off. He delighted in pissing her off. It kept things interesting. Boredom was a real risk, staying with her in the middle of a swamp. But this was where he had to be to get what he wanted.

"Before long, we'll have their defenses down and be able to walk in and find the girl and her friends," she mused. "If indeed I am right."

She shook her head. Of course, she was right. It made sense they would be with one of The Ancients. And each one was in a far-flung place, save two. Besides, she would love to see Esme or Isaiah again. Her smile turned wicked. It had been too long.

Siren stretched and yawned at her feet. She slunk out of the kitchen to the library to lie in front of the fire. Damien watched the cat with a wary eye. Infernal mates or not, he would never trust the beast. He didn't even trust Calliope.

Choosing to let sleeping beasts lie, Damien went outside. The moon was full again. According to Calliope, as the moon waned over the next week, it would lend more power to her spells, and they would see the wards break even faster. Damien thought the whole thing about the moon and witches was horseshit, but one thing about her he did trust was her abilities. He almost admired her for it. So, he didn't broadcast his low opinion of witchcraft.

The moonlight illuminated a form moving toward the house. Damien sniffed the air. Human, with an intoxicating element of fear. He smiled. Some desperate soul was coming to visit. Since their mating, a handful of wretches had come calling. He loved watching them tremble and stammer their deep desires to The Crone, begging her to help them. Those she couldn't or wouldn't help, she left to Damien to contract. Between the two of them, there had been a decent amount of business, and every fool left thinking they had gotten what they wanted.

And they did get what they wanted. There was just now a debt that would come due eventually. Whether the contract was with The Crone or a demon of the Courts was the only detail that differed from soul to soul.

He allowed the person to pass him, then turned to follow. Maybe his night wouldn't be boring after all. Damien froze when he heard a twig snap in the cover of the marsh. He lifted his head and sniffed again.

Multiple humans.

An ambush.

"Deodamnatis," he swore. They didn't have time for this.

The woman raised a shaking hand and knocked on the door.

"Don't—" Damien began to call as Calliope opened the door.

The report of a gunshot echoed through the swamp, and Calliope stumbled back. The property was suddenly swarming with humans carrying weapons. Damien jumped the planes to appear beside Calliope, who was bleeding from a wound in her chest.

"What the—" the female began before raising the gun to shoot at him. He strode toward her, and she managed to get two shots off before he reached her., absorbing the bullets in his chest. He grabbed the gun, wrenching it from her hand and breaking her wrist in the process. She screamed into the night, now alight with gunfire. Two arrows whistled his way. He caught one, the other embedding itself in his shoulder.

He roared and tore the woman's head from her shoulders in a spray of blood. Furious, he tossed the corpse into the mud and slammed the door. Calliope sat at the kitchen table, mixing ingredients.

"Hand me that belladonna," she instructed Damien, her voice weak.

"Will you survive?"

She grimaced. "Not if you don't hand me that belladonna."

He grabbed the extract and handed it to her while she continued mixing. Bullets zinged and splintered the wood on the porch. Siren sat beside Calliope, her shoulder pressed against her thigh. Once the potion was mixed, Calliope said a short incantation and drank it down, finishing with a sigh.

A window shattered, spraying glass into the foyer. Damien grabbed the grenade and winged it back, smirking with satisfaction as it exploded to a chorus of shouts and screams.

“This is unacceptable,” Calliope said, her voice markedly stronger.

“I’d agree.”

She rose from the chair and raised her arms. The bloody dress she’d been wearing disappeared, replaced by head-to-toe leather. Her cat o’ nine tails appeared in her hand.

“Let’s teach them some respect.”

Damien found himself smiling unexpectedly. He adored war but had been feeling something strange since he saw Calliope take a bullet to her chest. It was like adrenaline but different. He shook his head. It was gone now.

He ripped the arrow out of his shoulder and opened a portal. Two hellhounds emerged, their drool hitting the floor with a hiss. Their backs stood waist-high to Calliope, and their eyes danced with infernal flames. She clapped her hands.

“Inspired, Mate.”

“Après vous,” he offered with a wave of his hand.

She curtsied and threw the door open wide. The hellhounds rushed past her, leaping into the foliage and trailing fire in their wake. They bayed and howled while screams of terror and pain rose from the bushes. A man shot out of the trees, and Calliope caught him with her whip. He screamed and fell to the pursuing hound. Another tried to run, and Damien took out his legs at the knees with a sweep of his arm.

"The hounds will dine well tonight," he observed with a chuckle.

They stood side by side on the porch in the moonlight, disabling or killing any human who passed their line of sight. The hounds fed on them all.

The whole skirmish, from the first gunshot to sending the hounds back whence they came, took less than twenty minutes. Damien and Calliope stood in the kitchen once again.

"It would seem my swamp adventure is coming to a close," Calliope mused, rearranging and replacing ingredients on their shelves.

"What is the next move? Are we going after the whole city next?" Damien's excitement surged at the idea of that much bloodshed.

"No. That would be a waste of time. Fun, but a waste of time."

Calliope was once again in a sundress, her leathers and whip stowed in the ether. She began pacing the kitchen.

"We'll move to another of my properties. I have a compound near the Canadian border which will serve nicely for what we need to do to destroy the girl. And whichever Ancient is protecting her," she added. "I have everything I need at my other property."

The hem of her dress crept up her silky thigh as she sat down, and Damien licked his lips.

The next morning, they headed west. Siren took the form of a large black house cat with startlingly green eyes to travel with them. The engine screamed beneath the hood as they flew up the highway.

"I like this car," Calliope purred, petting the dashboard.

"I do, too. You know what it's called? A Demon."

"That's cute," Calliope tittered.

Anyone passing them would see a beautiful couple on a road trip. No one would see The Crone and The Devil, mangling cop cars as they drove across the country. Damien brought her to his place in the Midwest, and she oohed and ahhed over his torture devices and accoutrements. The corpses of the two Mormon girls were still there, and Calliope got excited when he told her they were virgins. She picked them over, removing certain bones for spells later.

The Crone hadn't been out of the swamp in a long time and was enjoying herself immensely. Siren seemed to be enjoying herself, too. She left around dusk to hunt the woods near his place and returned with blood on her muzzle. When she snarled at him, Damien saw the blood in her teeth. He snarled back.

"Don't ruin the moment, you two," Calliope called from her position, crouched above the remains of the dead girl on the bed.

"Wouldn't dream of it," Damien responded, presenting his middle finger to the cat.

"These are such wonderful gifts." Calliope thought for a moment before saying, "I'm going to fashion you a weapon."

Despite himself, Damien felt a rush of excitement. A weapon forged by The Crone was a precious thing.

"That would be... tremendous."

"I know," she responded easily, finishing gathering what she needed.

Earlier that day, Damien had gone to the local market and brought back some essentials for the next couple of days. They dined on the patio overlooking the woods. Siren lay at Calliope's feet.

"With the waning full moon, I'll be able to continue the ward-breaking spell and begin forming your weapon."

"I'm grateful," Damien said with a small nod.

"As you should be. I will choose which type of weapon you will wield. You do not choose."

"Understood."

"Good."

She took a swallow of red wine and licked her lips with relish.

"This is a delicious wine. Where did you find it?"

"When I was traveling before, I stopped at a bistro and had dinner. They offered this wine, so I took a case. I enjoy it as well. The varietal is Charbono. Old vine Italian."

She nodded and took another sip. They sat quietly for a moment, listening to the ambient sounds of the night.

"It's time to get to work," Calliope announced, dropping her napkin on her plate. "I brought the ward-breaker spells with me. Come, Siren."

She rose, carrying her plate to the kitchen, and Damien watched her go. She took her wine glass with her, draining the last of it as she walked. He tapped his fingers on the table in thought. Something strange was happening. He was enjoying the company of Calliope almost as much as fucking her. He... didn't want her to leave.

Best to keep that little revelation to himself.

The Crone set up in the upstairs study, meticulously laying out her spells. She lit the fire in the fireplace so Siren could curl up in front of it while she worked. Shadows danced over the books and flickered along the walls. She set her circle and closed it with her athame. Satisfied everything was in order, she concentrated on the spells.

Picturing the faces of the two Ancients, Esme and Isaiah, she felt along the ether for the wards with their magical fingerprints. She bumped up against one of Isaiah's wards and smiled. She began the incantation in low tones and pushed.

Downstairs, Damien threw back another shot of bourbon. It wouldn't get him drunk, but he liked the burn as it made its way down his throat. He chewed on the idea of caring about a being other than himself. It disturbed him. He poured another shot and downed it.

He examined the feeling he had when he saw Calliope take a bullet to the chest. After much deliberation, he realized it had been fear. He shook his head. Impossible. He feared nothing. A voice in the back of his mind disagreed.

Suddenly furious, he threw the decanter of bourbon against the wall. It shattered into sparkling bits of glass that caught the light as they flew. This was unacceptable. He was willing to accept the infernal mating. He admired her abilities and her lust for carnage. He acknowledged that he needed her help and couldn't destroy the girl without her. But to be afraid for her safety? No.

Anger was better. Anger was familiar. He wrapped himself in it like a blanket.

The crash downstairs caused Siren to get up from her spot in front of the fire and pace slowly. The Crone paused in her work to communicate with the cat. Learning Damien was throwing a tantrum downstairs, disturbing her work, incensed her. Just as she was making progress on another of Isaiah's wards, she could feel the barrier flexing and beginning to crack. She debated whether or not to head downstairs and whip him until she tired.

She decided to let it go for now. She was having fun destroying the Ancient's wards, one by one. She petted Siren on the shoulder and indicated she lay down by the fire again. Calliope settled in her circle once again and tried to pick up where she left off.

It proved harder to reconnect to the wards than she'd anticipated. After several failed attempts, she finally found Isaiah's ward again. She wasn't surprised the damage she had caused earlier had repaired itself, but it was annoying. She began the incantation again in a low voice and pushed.

In the periphery of her attention, she heard the car fire up and peel out. Good. Maybe he'd find something or someone to distract him for a while and let her work in peace. Siren stretched out and huffed a sigh.

Damien was in a mood. He was distracted and angry and couldn't think of a single thing that would make him feel better. His usual hobbies just wouldn't cut it tonight. He'd driven through the city in the Demon for hours, obeying traffic laws and scanning the rabble, waiting for inspiration to strike. Nothing was forthcoming. He felt restless, irritable, and discontented.

A boxing gym caught his eye. Maybe letting someone try to beat him into unconsciousness would help. He wouldn't even fight back. He just wanted pain. But no mortal would be strong enough to give him what he was looking for. He thought of Calliope and her cat o' nine tails. That would give him the pain he needed and likely end in sex. He turned back toward the house.

In the hours he'd been gone, The Crone had successfully broken another of Isaiah's wards and was feeling pretty pleased with herself.

The hour was late, so she broke the circle and gave herself permission to relax by the fire with Siren. She walked downstairs, poured herself another glass of the charbono, and carried both bottle and glass with her to the study upstairs.

The fire was smoldering, so she added more firewood. Siren sat on her haunches and sighed in contentment as the firewood caught and the flames grew. Calliope settled into a rocking chair next to the fireplace, placed the bottle and glass on a side table, and rested her hand on Siren's shoulder.

"Just like home, huh, friend?" she mused.

Siren growled low and bumped her head against Calliope's thigh.

The front door crashed open, destroying her moment of peace. Immediately irate, she stood to greet Damien as he blew through the doorway of the study. Siren stood and snarled, a deep growl reverberating through her chest.

"You dare disturb me this way! You had better have a good reason for being such an insufferable prick."

Damien shed his human form and stood before her in his infernal form. She felt her sex pulse but held firm to her indignation.

"You forget yourself, woman. I am a demon of the Courts. You will respect me."

"You forget yourself, you half-horned wretch. I was created to destroy creatures such as yourself. Maybe I will tonight."

Her sundress transformed into leather, and her cat o' nine tails materialized in her hand. Siren roared. The Crone let fly her whip and caught Damien across the chest, rending his flesh. Siren pounced and caught a piece of his ribcage with her jaws, tearing into him with her teeth.

He backhanded the cat, twisting his torso. The Crone pressed her advantage and whipped him hard across his back. Siren recovered quickly and went for another pass, this time catching his thigh with her claws. For the next half-hour, Siren and The Crone tore into him, and Damien finally felt his restlessness at ease for the first time that day.

The next morning, rain pattered against the windows, and a chilled wind whistled through the woods. Calliope sang to herself while she packed her spells to take to her house in Seattle. Siren was hunting again, and Damien sat downstairs, brooding. There was no sign of the violence from the night before, save for some broken furniture and spilled wine. Damien was completely healed, aside from the horn still growing back from the first time Calliope had tried to kill him. He uncorked a bottle of bourbon and poured himself a drink.

Siren came through the front door just as Calliope descended the stairs, flanking him. He smiled to himself and stood, drink in hand.

“All is ready,” Calliope called from the landing.

Damien tossed back the bourbon.

~

"We need to be ready as soon as possible," Isaiah said around a mouthful of pancakes. "Another ward was broken last night."

“We noticed,” Kaden responded.

“How close are we to being ready?” I asked, sipping my coffee.

“Not close enough,” Malachi interjected from his position, leaning against the doorway.

“Well, I think we're making very good progress,” Isaiah said, turning to Malachi, who crossed his arms.

“If we were up against a normal demon and a witch, maybe. But this is another league entirely.”

“What do we need to do?” I asked, pushing my plate away.

“My ley line manipulation is like second nature now, and we've practiced our attacks a million times. What more can we do?”

“Good question,” Draco joined in.

“Our coordinated attacks are good, but I believe we are vulnerable individually.”

Malachi uncrossed his arms and sat at the kitchen table with us, pouring himself some coffee.

“Okay,” I said, “how do we address that?”

“Each of you will train with me individually until I'm certain you can hold your own against someone as powerful as I am. Because,” he continued, “that’s essentially what we're up against."

“Good idea!” Isaiah bounced a little in his chair.

“When do we start?” Draco asked, pushing his plate away.

“After breakfast,” Malachi replied.

“Okie dokie,” Kaden said, standing up. "I’ll take the first spot since I’m the one most likely to be kidnapped.”

“What are you talking about?” I chuckled, turning to face him.

"I'm your best friend," he explained. "The best friend is always the one being kidnapped. Besides, I'm the cutest."

"Weird logic, but okay," I said, stacking the breakfast plates. Isaiah had a spell in place that kept his kitchen tidy, but I liked having something to do. "You can go first."

"Meet me by the clearing when you're ready," Malachi instructed, leaving the kitchen with his coffee.

"Aye-aye!" Kaden saluted.

After breakfast, we all gathered at the clearing. Mischief decided to come with us to watch me practice flying. While Kaden and Malachi worked on some defensive maneuvers, I meditated in the circle of stones, allowing my magic and the ley line energy to play without any directives. Then I practiced using the river of energy to lift me and move me from one side to another. It was still a far cry from actual flight, but I got closer every time I practiced. I had to admit it was also pretty fun.

"You're weak at defending your left side." Malachi whacked Kaden's left ribs with a walking stick he'd found at Isaiah's.

"Copy that."

Kaden adjusted his stance slightly until his left side was better protected.

“Better,” Malachi remarked, tapping Kaden’s left arm. “You’ve been relying too much on your magic in a fight. Most of the demons you and Angelica used to hunt were magicless. Your hand-to-hand is sloppy.”

Kaden nodded and reset his fighting stance. I set myself back onto the ground. I stood, wanting to try being lifted up while standing this time around. I took a nervous breath and willed the river to pool around my feet and lift me. I got about a foot off the ground before I lost my nerve and dropped back to the grass.

“Oh, cool!” Kaden called, taking his eyes off Malachi who then punched him in the face.

“Concentrate.”

“You got it. Thank you for pulling your punch.”

“I’m always pulling my punches, mortal.”

Kaden blinked.

“Did you just trash talk a little, or did you hit me harder than I thought?”

Malachi gave Kaden a ghost of a smile and indicated Kaden raises his fists again. They sparred for another twenty minutes while I practiced using the ley line to lift myself off the ground while standing. The biggest challenge was balance. I kept falling over sideways.

"Try making a bigger platform to stand on," Draco suggested from his spot next to Isaiah. They were deep in conversation about runework while Mischief lay sunning herself, eyes on me.

"Good idea," I answered, mentally kicking myself for not thinking of that earlier.

I widened the area of energy beneath me, and balancing was far easier.

"There you go! Look at you!" I could hear the pride in Draco's voice.

You're doing it! Mischief got up to trot over to me.

I tried getting the river to push me forward and nearly fell again, righting myself just in time. Slowly and steadily, I inched forward toward where Draco and Isaiah sat outside the stones. Kaden gave up focusing on Malachi and just blatantly watched me. Malachi dropped his hands and watched with him.

I was nearly to Draco, who was now standing, ready to catch me. I made it within his reach and let the energy go. Draco caught me with a laugh.

"I did it!" I couldn't believe it. Draco kissed me hard and looked at me. I'd never seen him so proud.

Mama, you flew! Mischief trotted back and forth between Draco and me, her tail twitching in the air.

"That was so cool!" Kaden clapped me on the shoulder before jogging to get a water bottle.

I wanna try. Can you take me with you?

"Sure!" I said, flush with my accomplishment.

Mischief and I made our way over to the circle of stones. She waited for me to get settled before climbing in my lap.

"Okay, little one. Hang on."

The river came to greet my magic instantly and pooled beneath me. I didn't want to get too far off the ground with my precious cargo, so I floated up just a few feet.

Can we go over to Papa?

"I'll try."

Slowly, we made our way over to Draco. At the last moment, Mischief leaped from my lap to his waiting arms.

That was fun!

"It sure was, my love," I laughed.

"Very good work, young one." Isaiah looked almost as proud as Draco. It still threw me when he called me 'young one' whilst looking like a teenage boy.

"A useful skill to have." Malachi joined us in the clearing. "Okay, who's next?"

Draco raised his hand.

"Excellent. Let's go."

Kaden made no effort to hide his relief at being done for now, flopping onto the grass with his water bottle. I sat next to him, Mischief trailing behind. The sun was warm, and the grass felt soft against my hands. I petted Mischief as she hunted a caterpillar. In this idyllic moment, I almost forgot the deadly seriousness of why we were here.

Draco fared much better than Kaden, and he and Malachi danced around each other, striking and retreating in a blur of motion. Draco paused to take his shirt off, and my mouth watered. He looked so beautiful. I felt a swell of pride mixed with lust, watching him hold his own against Malachi.

"Jeez..." Kaden said out of the corner of his mouth, "Take a cold shower, why don't you?"

"Get your empathic self out of my noodle, you perv," I said, throwing a water bottle at him. He ducked, laughing.

"It's not voyeurism if the emotions are broadcasted like a freaking bullhorn."

I rolled my eyes and watched Mischief pounce on something in the grass. She lifted her paw, found nothing, then lifted and looked at another empty paw.

"Very acceptable."

Malachi's voice drew me from my reverie. He and Draco stood in the grass, facing each other. Malachi didn't seem to have a hair out of place; Draco was glistening with sweat and winded.

"That was fun," Draco said, clapping Malachi on the shoulder. Malachi almost smiled.

Draco raised a hand, and I threw him a water bottle. Our perfect moment of peace shattered with the resounding crack of another ward falling. Isaiah's smile drained from his face.

"Damn," he said softly, sighing. "I need to make a phone call."

Isaiah rose from the grass and made his way back to the house, brushing at his robes as he walked.

"Angelica," Malachi beckoned me over.

I stood and walked to him and Draco.

"Draco is fine. Why don't you pair off and spar? I'll continue working with Kaden."

"Sounds good, boss!" Kaden called from his spot next to Mischief. He got up and joined us in the clearing. Draco and I moved about thirty yards from Malachi and Kaden and squared off. Mischief took the opportunity to return to the house. She never liked watching Draco and me spar. It made her nervous.

"Alright, Winter. It's on."

"Whatever you say, Angelica."

We sparred for about an hour, with Malachi interjecting instruction here and there. It felt amazing to be physically fighting after so much time spent on the energy of ley lines and magic. Draco was a better fighter than I, but I managed to hold my own with minimal comments from Malachi.

The plan involved me being mostly protected, so, hopefully, I wouldn't have to do much hand-to-hand combat when the battle was on. But it is better to be prepared than beaten, captured, or killed.

Did you beat him up, Mama? Mischief asked when we made our way to the kitchen, starving from our exertion.

"No, sweet girl. We would never hurt each other."

I know. I was trying to be supportive.

Draco laughed, and I picked her up to give her a hug. She squirmed in my arms and gave me a straight paw to the face. I set her down before she tried to leap, and she wobbled off to find a sunbeam. I sat in one of the kitchen chairs and pulled my plate of meatloaf toward me.

"I love this kitchen," I declared around a mouthful of mashed potatoes. I hadn't even realized I was in the mood for meatloaf and mashed potatoes until it was in front of me.

"It is pretty excellent," Kaden agreed, digging into his lo mein.

A companionable silence fell over the table as we concentrated on our food.

~

Calliope and Damien arrived around midnight. Gravel popped under the tires as the car pulled up to the front of her house. They sat for a moment in the circular driveway before opening the car doors. Siren hopped out and returned to her panther form, striding up the front steps toward the door. She took a right at the front of the house and slunk around to the side toward the back, where a large cat door was built into the back.

Calliope exited the car and strode up the front steps to her front door. Damien watched her disarm the wards so he would be allowed to enter. He stepped out of the car and met her on the porch.

"You may enter now," she declared, opening the door and walking in before him. He followed her into the foyer and looked around.

The place was full of marble and cold as a museum. Heavy drapes blocked the moonlight, and tapestries blanketed the walls. A massive fireplace was recessed into the far wall, and Damien smirked when The Crone brought it to life with a quick rune. Siren padded into the room and, ignoring Damien, curled up on a huge, soft blanket in front of the fire. Curios and oddities peppered the space, some under glass, some not. Damien contemplated a rat skull on a pedestal under leaded glass with raised eyebrows.

"It's radioactive," Calliope provided.

"I see."

Damien moved on to inspect the rest of the room. Books were everywhere—in bookcases, in towers on the desk, in mounds on the chairs, and in piles on the floor. A lifetime of dust covered everything.

"Did you never consider having a caretaker for this place?" Damien queried, running his finger through the dust along the spine of a red leather book.

"I had one. He died, and I never replaced him."

"Mortals."

She nodded and settled into the rocking chair next to the fireplace. Siren slunk up to her and sat next to the chair, allowing Calliope to place her hand on the back of her neck. Calliope closed her eyes, and a wind picked up in the room. Damien watched as the dust was removed from the red leather book he'd been touching. The ambient light in the room brightened as grime was removed from the lightbulbs. Damien wondered for a moment that the lightbulbs still worked but held his tongue. Soon, the place sparkled, and a fresh scent replaced the smell of neglect.

"A neat trick," Damien said, sitting on a chaise on the other side of the room from Calliope.

"Useful."

An awkward silence followed. Damien realized they had nothing to say to each other when they weren't fighting or working toward the destruction of the witch. Calliope seemed unperturbed by it, but he decided to make himself scarce. He thought he might go out and find some trouble to get into.

~

I spent hours every day sparring with the boys and working with the ley lines, rewarding myself by practicing flying. Mischief always showed up to watch, and I often took her up with me, to her abject delight.

A ward was falling every three or four days, and Isaiah told us there were only a handful left before we were exposed. All we could do was prepare. The imminent conflict was drawing closer every day. It was unsettling.

I took one more stab at putting some speed behind my flying and managed to avoid falling, which was a win. I was getting the hang of the balance, which was sort of like surfing. I fell a lot. I had never been great at surfing.

"You look so cool up there," Kaden remarked, pausing in his sparring with Draco to grab a towel.

"Thanks. It's pretty fun when I'm not falling."

"I can only imagine. Totes jealous."

He wiped his body down and took a drink of water before gesturing to Draco to continue.

I had Mischief nestled in my arms as I took flight once more. She purred softly as I gained altitude, and the sounds of the boys grew faint.

I like flying, mama. It makes me feel all light inside.

"I hear you, baby girl. I feel the same way."

The air was crisp, and the sky was gray. Clouds gathered on the horizon, threatening snow. The smallest droplets of rain speckled my face and caught in Mischief's fur. I decided this was as good a time as any to descend. I made my way gently to the ground and placed Mischief in the grass.

A headache had been behind my eyes all day, and I had been combating it with copious volumes of water, sure it was dehydration. Suddenly, the ache began to grow, causing my vision to tunnel. I stumbled and sat down in the clearing with a groan. The pain was now so intense I could barely see. I tried to call for Draco or Kaden, but my voice came out as a whisper.

I lay in the grass in a fetal position, hugging my knees to my chest in agony. Then everything went black. When I came to, the whole menagerie was circled around me. Kaden and Draco looked worried, Isaiah looked curious, Mischief was on my chest, and Malachi looked diffident, as usual.

"There she is," Isaiah said cheerfully.

"What happened?" I asked, pinching the bridge of my nose.

"The current Gatekeeper has passed from this realm. The responsibility is now yours," Malachi said, squatting down next to me. "Do you notice anything different?"

I started to say no, that I was fine when I looked around. I could see *everything*. The energies crossing each other in the circle, Malachi's aura, the river of the ley line pulsing beneath me, louder than a drum. I chanced a glance at Draco and could see his aura, which was so beautiful it took my breath away.

"I can see… everything."

Malachi nodded.

"Now it's time to really start learning what it means to be a Gatekeeper."

Everyone filed back to the house, leaving Malachi and me alone in the clearing alone.

"The first thing we're going to work on is being able to shut off your Vision."

"Thank the gods. I was worried this was going to be a twenty-four-hour thing."

"It will be," he answered, "but you'll learn techniques to mute the colors and sounds. Sort of like a new vampire."

"That makes sense."

"First, find one thing to focus on and try to filter out the rest. It's difficult at first, but you'll get the hang of it."

His unsolicited support took me a moment to process. I had grown accustomed to his brusque manner of doling out crumbs of encouragement.

"Okay."

Seeing as the ley line was the most obvious, I chose to focus on that. I concentrated on the color and the movement, the rushing sound of the river of energy.

"Now try to bring your focus to a pinpoint, allowing no other stimuli in."

I tried and tried, but I could not shut off the noise and color around me. After a half-hour of intense concentration, my forehead beaded with sweat, Malachi called a time-out.

"Let's try this another way." He sat down in the grass with me. "I'm going to keep talking while you concentrate. Try to focus on only two stimuli. It's a bit like playing the piano—one hand works on the melody while the other complements it. Keep your eyes open."

I obeyed, and he started talking. I watched the ley line shimmer and rush, trying to block out everything else except that river of energy and Malachi's voice. Slowly, the surrounding miasma of color and sound began to ebb. I wanted to tell Malachi it was working, but I didn't dare risk disrupting my concentration. I breathed into it, and more fell away.

"Very good work."

I smiled and broke the connection. The world rushed back in.

"You've gotten very proficient at working with ley line energy while fighting. Now, I want you to concentrate only on what we just practiced. Try it at all times during the day until you can turn it on and off like you can with the ley line energy."

"Thank you, Malachi."

"No need to thank me, Angelica. You've made a lot of progress in the short time you've been working on this. It is your birthright and destiny."

I resisted the urge to say 'DUM-DUM-DUMMM,' but it was a close thing. Maybe I was a little punch drunk.

"We'll stop here tonight."

"Tonight?" I looked around and realized the sun had set. "How long were we out here?"

"Four hours."

"Holy shit. With my new 'Vision,' as you call it, the delineation between day and night is not as stark as it used to be. Is that normal?"

"I can only imagine. I cannot speak from personal experience."

I turned to walk toward the house and nearly fell. Being able to see the energy of everything was pretty disorienting. I concentrated on my feet and practiced shutting out everything else. It was slow going back to the house, but Malachi patiently kept pace with me, steadying me every time I stumbled.

"It's like watching a baby giraffe learn to walk!" Kaden called from the front porch.

"Har-dee-har-har." I rolled my eyes.

"I've got it from here, Malachi," Draco said, lifting me. I squealed as he threw me over his shoulder.

"Not so fast! It'll make me nauseous!"

"I got you, baby."

Draco carried me all the way to my room, gently laying me down on the bed. He looked so beautiful. His aura was nearly indescribable, a clean and bright heart-wrenching shade of blue shot through with silver threads. If I concentrated, I could see his heart pumping and the blood whooshing through his veins. I focused on shutting out everything but him. It was coming a little easier every time.

"I have an idea." I sat up and held my arms out to him. He stepped between them, nudging my knees open to get closer.

"Oh? Pray tell." He kissed the top of my head, and I kissed his stomach.

"It's better if I show you."

The snow began to fall.The next morning, I nearly fell out of bed when I opened my eyes. In my sleep, I had forgotten about my new Vision.

"You okay, baby?" Draco asked, muffled by the blanket.

"Oh, I'm fine. Just adjusting to my new reality."

"Describe what it's like," Draco said, sitting in bed beside me.

"It's like a mix between a kaleidoscope and an x-ray. I can see the energy of everything in sight, all in different colors and pulses. If that makes sense."

"Sounds a bit dizzying."

"It is. Definitely going to take some getting used to."

I took the opportunity to practice right away. I focused on Draco and allowed the rest of the world to fall away. Shifting my sight from him to the door, my Vision came with me.

"Mischief is coming."

"How do you know that?"

"I can see her on the other side of the wall, headed toward the door."

"Wow. What a gift."

"I'm choosing to see it that way. Without Malachi's tutelage, I see how this could be really awful."

Hey, guys.

"Hi, sweet girl. Have you had a nice morning so far?" I patted the spot next to me on the bed, and she climbed up. I stroked her downy head, and she purred softly.

I went for a walk around the house. I feel another ward is going to fall soon.

"That's not great," Draco remarked, running his hand through his tousled hair.

"We'd better let everyone know." I started to get up, and Mischief stopped me with a paw to the thigh.

I already told them. They're talking in the library.

"Then a shower is in order." I winked at Draco. He smiled.

"Do you need help showering with your new Vision?"

"I might."

You guys are so obvious. I'm not a kitten. I'll see myself out. She bumped her head against my hand and hopped to the floor.

Draco started the shower while I got undressed. He was testing the temperature with his hand when I walked in and rubbed his back.

"You're wearing far too many clothes to shower, honey."

"I know," he responded, turning and straightening up to encircle me in his arms.

He ran a hand along my bare shoulder and down my arm, creating a trail of goosebumps. Everywhere he touched, he followed up with a kiss. I let out a contented sigh and surrendered to the sensation of his touch.

"Mmmm… That feels nice."

“What about this?” Draco said, trailing kisses along my collarbone.

“Mm-hmm.”

He kissed his way to my neck, and I tilted my head to allow access. His hands explored my body while I tugged at his sweater. He broke away long enough to remove his sweater and undershirt, leaving him shirtless and gorgeous. He lifted me by my thighs, and I laughed, wrapping my legs around his waist.

“You’re still wearing too many clothes,” I pointed out, running my fingers along the top seam of his jeans and tugging at the zipper.

“Not yet.”

He set me down in the stream of warm water and grabbed the shampoo. He poured the peppermint-scented gel into his hands and began massaging it into my hair. I closed my eyes in bliss. After washing and conditioning my hair, he picked up the loofah and worked the shower gel into a lather. He started with my shoulders, slowly rubbing and washing every inch of my arms and breasts. He turned me and massaged my back, cleaning and rinsing down to my waist.

“Lower, please,” I begged.

He huffed a laugh and reached around to pet my sex, gently using his fingertips to tease my folds. He brushed against my clit, and I moaned, desperate for more. He turned me back around and dropped to his knees. He parted my sex and laved his tongue over every sensitive inch. Slowly, firmly, he pressed a finger into my channel, flicking my clit with his tongue. I gasped and threw my head back. An orgasm was building, and I panted, watching the colors swirl behind my eyelids. Draco groaned, and it sent me over the cliff. Waves of pleasure crashed over me again and again, making my knees buckle. Draco steadied me by my waist and stood. He kissed me, and I could taste myself on his lips.

I grabbed him and pulled him to me by his belt with a grin. He chuckled and brushed my hair back. I unbuckled him and pulled his jeans down.

"Your turn."

"If you insist"

Later, we all convened in the library.

"There are only two wards left intact at this point," Isaiah stated, his brow creased.

"So, what do we do?" Draco asked from the couch where we were sitting.

"All we can do is prepare. A war is coming," Isaiah responded. "I suspect by this time next week, we'll be in battle."

"That's not great." Kaden grabbed a bagel and flopped into a chaise.

"Keep your spirits up. You are better at this than I predicted," Malachi added from the doorway. He turned to me.

"You've been practicing?"

"Every second of the day."

"Good."

He scratched his chin. "I believe we are as prepared as we're going to be. Take this time to make your peace and get your affairs in order."

"What the devil?" Kaden exclaimed. "I thought you just said we were good at this. Now you're telling us to prepare for death?"

"All warriors go into battle at peace with death. If you hesitate due to mortal fear, you will die."

That last statement was met with silence. Kaden, Draco, and I exchanged glances. Draco took my hand and kissed it. The bottom line was Malachi was right. If we feared this coming battle, we would lose. We had to face it with no reservations.

One of the last wards fell with an earth-shattering crack.

~

"Isaiah and Esme have just a couple of wards left apiece," The Crone announced.

"That is welcome news," Damien responded.

Damien was in a funk. He had tried to contact Legion to no avail. His powers were almost back to full strength, but he felt no urge to go out and spread chaos. The only time he felt anything close to pleasure was when Calliope was either fucking him or fighting him. He awoke thinking of her. He coveted her. She had become… important to him.

"I give it one more week, and we'll be able to just walk in and do as we please."

Calliope smiled through her hair at Damien. She was pleased with herself. Damien smiled back.

"Shall we celebrate?" she asked with a coy look.

Damien allowed his true form to manifest, tearing through his clothing. Calliope's pupils dilated. He stood and made his way to her like a predator cornering prey.

"Let's." He fisted his clawed hand into her hair and kissed her hard. She giggled into his mouth. Damien devoured her like a starving man. Calliope raked her nails down his arms.

"Worship me," he growled.

"Worship me," she responded, digging her nails into his skin hard enough to draw blood.

He lifted her by her thighs and slammed her against the wall, cracking the plaster.

"Worship me!"

She laughed and slapped him across the face. With a twist of her wrist, her clothing disappeared.

"Only power is worthy of my worship."

"I *am* power."

"You are powerful."

He roared and speared her with his cock. She threw her head back and moaned. He thrust into her, desperate to consume her, to own her, to make her worship him and never leave. Her moaning turned to screams as her sex tightened around him. He fucked her harder, punishing her for not obeying him. She came with a cry, and he followed, filling her with his seed.

She climbed down from his waist and patted him on the cheek. The patronizing gesture infuriated him. Sensing his rising fury, she shot him a warning look before turning on her heel. As she walked away, she twisted her hand in the air and was once again clothed. He watched her walk away with a sinking feeling.

Later, they sat on the porch overlooking the Pacific.

"Isaiah's location is closer to us. I recommend we start there." Calliope swirled her wine, inspecting its color in the light.

"That makes sense."

"Just a few more days, and the Ancients will be exposed."

Damien felt a frisson of excitement cut through his malaise. He began to remember why he sought out The Crone in the first place. He had been distracted by her over the last month or so. He had briefly wondered whether she had cast some spell on him to cause these feelings, but he dismissed the idea. The depressing truth was he was a demon in love.

"Shall we walk in and raze the place to the ground? Or were you thinking something more subtle?" Calliope asked, standing from the table with wine in hand.

"I intend to have as much fun as possible. Even if the little witch and her friends aren't there, it would still be something to destroy an Ancient."

The Crone laughed.

"You destroy an Ancient? You had better leave that to me."

Damien stood from the table as well. His voice took on a deeper timbre, and fire danced in his eyes.

"You think me incapable?"

"I think you're very good at what you do—torturing, raping, sowing chaos. Taking on the power of an Ancient is another level."

Damien seethed. Her condescension got under his skin in the worst way. She insulted him twofold: his power and his love for her. She didn't think him worthy. He began to manifest.

"Don't threaten me with a good time, demon. I'm simply stating the facts. If you want to succeed, you need to let me handle Isaiah or Esme. Don't let your pride ruin your chance for success."

Damien stalked off into the house. He shifted back into his other form and decided to go out. He headed to the closet and picked out his sharpest outfit. If The Crone didn't appreciate him, he'd find someone who would.

~

Isaiah had announced one ward remained, and we were rapidly approaching a battle, so I spent every second of the day practicing using my Vision and manipulating ley lines. Malachi had suggested I practice finding them wherever I was, and I found them everywhere. It was comforting to feel the pulse of energy wherever I went.

Kaden was a foundling, and Draco's parents were dead with no siblings, so "getting our affairs in order" basically meant telling each other, "I love you." I sure as hell wasn't going to reach out to my parents to say "goodbye." That would be madness. Being in a place for the last few months where they couldn't find me had taken a lot of stress off my shoulders as it was.

I hadn't touched a tarot deck in months. Part of me missed it; another part didn't need the reminder that a battle with power hitters like The Crone and Damien was coming. Realizing I hadn't done a reading in so long, I pulled my tarot deck from my desk earlier that morning, but my Vision made it nearly impossible to read the cards. So there was that.

Kaden, Draco, and I were sitting out on the patio having lunch when Isaiah swept in to join us.

"I have a feeling they are going to come to us," he said, spearing an olive from a bowl.

"What makes you think that?" I asked, taking a bite of my sandwich.

"Calliope has spent a lot of time and energy disabling and destroying the wards around my and Esme's places. Why do that if she didn't want to just walk right in?"

"That's a really good point." Draco ran a hand over his face.

"So that will give us home court advantage. That's a good thing, right?" Kaden sat back with a handful of baby carrots.

"Yes, that's true," Isaiah responded. "It gives us a slight advantage, which we can use for sure."

"Do you think Esme will join us if we're attacked first? And should we rush over there if she's attacked?" I had been so focused on our homestead that it hadn't occurred to me earlier what we would do if Esme were attacked.

"If Esme wants my help, I'll go. As much as I would welcome the help, I am not depending on her coming here." Isaiah chewed a cucumber thoughtfully. "There are so few of us left. I wouldn't expect her to put her life on the line like that. But I don't know for sure. Hm. I need to make a phone call."

Isaiah grabbed a sandwich from the plate in the middle of the table and swept back into the house. Kaden pensively regarded the doorway through which Isaiah had gone.

"I wonder why he uses the little kid form. His adult form is hot."

"I've wondered that myself," I said, turning to consider the doorway with him. "There has to be a story there."

"We may never know." Draco picked up another sandwich from the plate.

Isaiah dialed Esme's number. It rang a half-dozen times before she picked up, a little breathless.

"You know, I have a cell phone, Isaiah."

"Yes, but I like the tactile feel of an old phone."

"Gotcha. So, what's up?"

"I was discussing the coming threats with the children and wanted to ask whether you want my help if Calliope comes to you."

"That's sweet of you to offer. Maybe."

"Maybe?"

"I don't think she'll hit my place first."

"No? Why not?"

"Because she has a house in Seattle."

"Oh. Interesting."

Pause.

"Esme?"

"Yes, Isaiah?"

"Would you help me? One of the kids they're after is the Gatekeeper. I fear I may not be enough to keep her safe."

Esme was quiet as she considered her next words.

"Calliope and Damien?"

"Yes."

She sighed.

"Yes, I'll be there in a day or two. Let me get a few things in order here, and I'll take a portal."

"I really appreciate this, Esme. I know this is not ideal."

"Far from it. But there are so few of us left. We need to stick together."

"I'll see you soon," Isaiah said with a smile.

"See you in a couple of days."

They disconnected, and Esme placed the receiver in the cradle. She kept the phone only for Isaiah's calls.

The last time she had seen him was shortly after Calliope had murdered Agatha. None of them was the same. The shock of realizing they could be killed was enough, but it had been by Calliope, one of their own. Esme and Isaiah were betrothed when Agatha was killed. Isaiah became obsessed with trying to bring Agatha back and with finding and bringing Calliope to justice. Neither happened and after two years of watching him kill himself over the obsession, Esme left.

It shattered her. She traveled around North Africa and the Fertile Crescent for a time, slowly making her way north through the Mediterranean countries. She watched helplessly from a distance as the Ancients were slaughtered, one by one, by Calliope. Each death took a piece of her heart, and she wept for them all. It wasn't until relatively recently she felt whole enough to return to the States. The threat to Isaiah felt different.

She walked into the bathroom to run the shower. Isaiah had hurt her with his neglect at the time, but, as he'd told her over the years, he still loved her. And she loved him. Eternally. She had known, at some point, they would reunite. The only thing which had kept her from heading to Canada until now had been pride. She shook her head. Silly thing.

Seventy five years ago, Isaiah built them a chalet in BC atop an intersection of ley lines. He constructed the foundation with runed Italian stonework, laying the stones himself over a period of six years. When it was finished, he asked her to come home, and when she wouldn't return from Africa, he razed it to the ground in grief.

He had been using the form of an adolescent for decades, too heartbroken to see himself as she once saw him. It was a logic which escaped her understanding, but there it was.

She undressed and stepped under the stream of hot water, closing her eyes as it sluiced over her head and neck, wending its way over her body. She washed her hair and thought of him. She washed her breasts and stomach and thought of him. Now that she was to see him after centuries, she could think of nothing else. When she had said 'yes' to helping, she hadn't fully thought it through. Every molecule in her being wanted to help him, so she'd said 'yes.' Now, under the falling water, she realized that meant they would see each other.

"Oh, shit… What am I going to wear?"

She packed carefully, as if Isaiah wouldn't be able to provide anything under the sun for her if she wished. The only piece of jewelry she had worn for centuries was a simple silver locket Isaiah had given her when they were courting. She clasped the locket around her neck, and the small oval nestled in the hollow of her throat. She noticed the pulse in her neck feathering under the skin.

She tidied the house and locked the windows. The sun was warm on her back as she locked the front door and placed the key under her neighbor's mat. The thought popped into her brain that she might not be coming back. It was met with no trepidation, no fear. Isaiah needed her.

Isaiah hung up the phone with a wide smile. Esme. Coming home. The circumstances were less than ideal, but he would see her again. Even the threat of The Crone and Damien could not dim his excitement. He was still smiling when he allowed his true form to return. The monk robes disappeared and a finely tailored shark gray suit took their place. He wore a charcoal shirt without a tie underneath the jacket and ebony cufflinks. He ran his fingers through his hair, then gave the wall mirror a wink, straightening his cuffs.

"So, children, I have news."

Isaiah strode back onto the patio and took a seat at the head of the table. He was met with wide-eyed shock. My jaw actually dropped. The voice was the same, but instead of an adolescent boy in flowing robes, there sat an amazing specimen of a man. His violet eyes twinkled with mirth as he crossed his ankle over his knee. Polished Louboutins drew my eye. He was beyond gorgeous. I turned to find the same shocked expressions on the faces of my compatriots and slowly shook my head.

"What the hell, man??" Kaden was the first to break the stunned silence.

"Esme has agreed to help us," Isaiah said, acting as if everything were normal. Then he laughed. "I know this is a shock, but I can't very well see Esme as a fifteen year-old boy, can I now?"

His face split into a heart melting, panty-dropping smile as his news slowly sank into my befuddled brain.

"Wait, Esme is coming?" I felt a tickle of excitement. "She's going to help us?"

"She will be here in the next couple of days. Maybe sooner." He grinned again, and my heart skipped a beat.

"Well, that's great news, gorgeous!" Kaden exclaimed slapping the table. Isaiah chuckled.

"I am excited for you to meet her despite the circumstances. She's an incredible woman."

"Another Ancient. That really is good news, Isaiah. Thank you." Draco smiled and brushed my hair over my shoulder.

"You can close your mouth now, sweetheart. It would seem this one's heart belongs to another."

"Oh, ha. Thanks for that," I responded "Though there may actually be drool…" I said partly to myself and checking my face. Draco laughed.

"She does own my heart, children. She has for centuries. And now she's coming home. I don't care about the doom and gloom today. We will welcome her with a party!"

"With all… five of us?" Kaden asked with a raised eyebrow.

"Six, if you count the guest of honor," I added.

"It'll be perfect. I'll let Malachi know." Isaiah stood from the table and practically skipped back into the house.

A beat passed before anyone spoke again.

"That picture did not do him justice," I said with a low whistle.

"Seriously. What does that say about whom we're about to meet? She's going to be beyond stunning." Kaden took a sip of soda water and sat back.

"I'm just thrilled we have another Ancient on our team," Draco added. "We can use all the help we can get."

"Is she the redhead? I'm trying to remember which one she was. I want to be prepared," Kaden said with a thoughtful expression.

"How do you prepare for perfection?" I asked. "I fear losing my concentration if Malachi manifests his true form during the fight."

"And how did we get here? Angels and demons, Ancients and Gatekeepers… It's a little surreal," Kaden remarked. "Just a few months ago, we were hunting a slime demon in the sewers."

"Correction, *I* was hunting a slime demon. You were too afraid of botulism."

"E-coli. Not the point."

"No, I get you. It is unbelievable. But here we are."

"Here we are," Draco echoed softly.

Every one of us flinched when the final ward failed with a resounding crack.

The Crone smiled. It was time. Damien was sitting in a chair, brooding. Let him brood. She now had an opportunity to take out an Ancient, possibly two. This was a fine day.

“The wards are down,” she announced.

Damien looked up at her, a wide smile splitting his face.

“Are you serious? Or is this another fucking game?”

She shook her head.

“No game. It’s time to have a little fun.”

He stood and strode over to her, grabbing her by the waist. She giggled as he threw her over his shoulder.

“Oh, help me! A big nasty demon has me!” she joked in a high-pitched voice, swatting at him with soft slaps to his back. He chuckled and squeezed her ass with one hand, the other arm a band across the backs of her thighs.

He carried her up the stairs to the bedroom and threw her onto the pillows with a salacious grin. She writhed in anticipation.

“Show me your true form,” she demanded.

He calmly unbuttoned his shirt and folded it neatly on the chair.

“I’m going to make you beg for release, my little witch.”

She let out a shuddering breath.

They stayed one more night in the house and left early the next morning.

Esme trekked to a secluded spot in a nearby woods. Checking to make sure there were no witnesses, she opened a portal and, with a nervous breath, stepped through.

Isaiah was a man on fire. He had the kitchen produce enough food to feed a small army. The house was decorated with fairy lights and floating orbs of warm light. He swept in and out of the room, adjusting small things here and there.

"I think it's perfect, Isaiah," I offered.

"Yes, but I want it to be perfect."

Kaden and I exchanged glances.

"Okay, man. Do you need anything?" Kaden asked.

Instead of answering, Isaiah rushed into the kitchen to check the food again.

"He could use a drink," Kaden said and let out a breath.

"So could I, honestly," I answered, making my way to the bar. Kaden followed me.

"I'll get it," Kaden offered, making his way behind the bar. "What'll you have?" He asked, throwing a bar towel over his shoulder.

"I'll have a bourbon and soda, please," I responded with a smile.

"One bourbon and soda, coming right up."

He poured my drink and slid it across the polished bar.

"And the bartender will have a beer."

"A fine choice."

"Oh, are we having drinks?" Draco asked as he came in, carrying a tray of shumai and salted edamame.

"I'm surprised Isaiah is letting you help," I said, indicating the tray in his hand.

"It took some cajoling, for sure," he said with a smile. His smile deepened when he took me in. He set the tray down and let out a low whistle.

"You look beautiful, my love." He turned to Kaden and said, "I'll have a beer as well, barkeep."

Kaden acknowledged the request with a nod.

"You look amazing yourself," I said, appreciating the sight of my man in a suit. "You sure clean up fine."

Draco shot me a dazzling smile, which stole my breath for a moment. He leaned down to kiss me, and I snaked my arms around his waist.

"Ahem," Kaden interrupted, "are you blind? Missing the best looking of the group…" Kaden ran his hands through his hair and pooched his lips out in a mockery of runway models.

Draco broke our kiss and said, "You look beautiful, too, my man," and clapped Kaden on the shoulder.

"I'll admit I thought Isaiah mad when he suggested a party, but this is fun so far. Dressing up is fun." I took a sip of my drink. "Plus, I like seeing my boys all pretty."

"I'll have what she's having," Isaiah said as he joined us. "Everything is as perfect as it's going to be. I may as well have a drink. I'm all nervous."

Kaden fixed Isaiah's drink, and we stood in comfortable silence for a moment. Isaiah's hand froze midair. He paled, and his eyes went wide.

"She's here."

He placed his untouched drink on the bar and strode quickly from the room.

"Do we follow him?" Kaden asked.

"I don't know," I answered. "They haven't seen each other in a long time. Maybe we should give them some privacy."

"Agreed," Draco said, setting his beer on the bar.

Esme managed to keep on her feet when the portal spat her out onto the grass, but it was a near thing.

"Portals are not the most graceful way to travel," she said under her breath. She looked herself over to make sure everything was in place and her outfit still looked perfect. There was a grass stain on her shoe. She began to reach down to rub it off when she heard him.

"Esme."

Esme froze. She slowly straightened, turned, and shattered into a million pieces. She let out a little gasp and stood straight. Seeing Isaiah was more of a shock than she could have ever anticipated. He looked like he'd stepped out of a recurring dream. Better.

For his part, Isaiah was practically in tears. She was so beautiful, his heart stuttered. She breathed his name.

"Isaiah."

She took a couple of hesitant steps toward him, and his stomach flipped over. He closed the distance and wrapped his arms around her. She, in turn, threw her arms around his neck. They held each other for forever and no time at all.

Isaiah pulled back and gazed into her face, his eyes stinging with tears. They were close enough that he could see his image mirrored in her eyes. Her lips parted, and she let out a small breath. He dipped his head toward her, slowly, until he could feel her breath on his cheek.

Then he kissed her, and the world made sense again. She sighed a moan into his mouth, her knees weak. He held her tighter and deepened the kiss. He trailed a hand up her side until he cupped her face. They clung to each other and kissed beneath a darkening sky. After a time, Isaiah pulled away.

"There's a party to welcome you."

"Really? How lovely."

"Come to the house and meet the kids. Malachi is here, too."

"It will be nice to see him again."

They turned toward the house, hand in hand.

“They’re coming!” Kaden announced from his perch at the window. “Wow…” He sighed, “What a beautiful couple.”

“Kaden, get away from the window,” Malachi said, taking a sip of his martini.

“Yes, Dad,” Kaden pouted. Malachi crossed his arms and leaned against the doorway.

“In his defense, this is the first time two Ancients have been together in a very long time,” I remarked. “Plus,” I added, “they are like royalty in the supernatural world. It’s kind of a big deal.”

Draco walked in with another tray of fancy snacks. “I don’t know who is going to eat all of this, but it sure looks good.” He plucked a mini quiche from the tray and popped it into his mouth. He nodded his approval.

The front door opened, and Esme and Isaiah strolled in. Malachi was the first to move from his post at the doorway to the kitchen. He walked up to Esme and enveloped her in a huge hug. She giggled.

“It’s nice to see you, Malachi,” she said, her voice muffled by his chest.

"The pleasure is all mine, Esme. It's been too long."

"It really has," she responded, pulling back to look at him with a beatific smile. "So," she said, looking over his shoulder at us, "these must be the kids I've heard so much about."

Kaden stepped forward with his hand out. "Kaden, ma'am. It's a pleasure to meet you."

Draco was next. "Thank you for coming to help. It means the world."

Her eyes met mine, and she gracefully closed the distance. "And you must be Angelica, The Gatekeeper."

"So I've been told, ma'am."

"No need for this 'ma'am' business. Please call me Esme. We are all just warriors here."

To call her stunning was an understatement. Her figure was willowy and perfect, and her red hair and green eyes shone. She inspected me and fingered a lock of white in my hair.

"What happened here?" she asked, turning to Isaiah.

"Demon magic attack. I was nearly drowned in a hotel bathtub," I answered. "That was before we met Isaiah."

She turned her attention back to me with a gentle smile. “So it didn’t happen when you were trapped by The Crone?”

“No, no physical effect from that particular nightmare.”

“You’ve been through a lot.”

“I have incredible people around me. They’ve all saved my life.”

“Well,” she said brightly and clapped her hands, “where’s the booze? I think it’s time to get this party started.”

She squeezed my shoulder as she walked past me to the bar. The room dissolved into casual conversation as she fixed herself a Manhattan. Once she had a drink in hand, Isaiah met her at the bar and retrieved his untouched bourbon and soda.

“This place is beautiful, Isaiah. And look at all the books!” Esme craned her neck backward, just as we had when we first entered his house.

“Here, take your drink with you and I’ll give you the tour.”

“I would love that.”

They left the room, and the energy level in the room dropped dramatically. I hadn’t realized how much passive power they had until they took it with them to the next room. It bolstered my confidence to have them on my side in all of this. I wasn't ready to hope we could defeat The Crone and Damien yet, but at least we were going to put up a hell of a fight.

"Wow…" Draco breathed.

"I know, right? You should see what they look like with my Vision. Absolutely breathtaking."

"You guys had better get over here and have some of these shumai. They are the best I have ever had," Kaden called from the bar.

Draco and I linked hands and made our way over to Kaden, who was popping dumplings into his mouth with relish.

When Isaiah and Esme returned from the tour, we sat down to dinner. The most succulent meats, cheeses, and pears began our fare, followed by plates of Wagyu beef and braised potatoes. We had a choice of lamb or duck with grilled vegetables and the best bread I'd ever tasted with churned butter. Plates piled with delectable desserts: cheesecakes, mousses, chocolates, and fruit tarts. I had never eaten better.

“That was delicious, Isaiah,” Esme commended.

“It was good, wasn’t it?” he asked with a pleased smile. “I’m so glad you enjoyed it. Now-” he stood and clapped his hands, “it’s time for games!”

“Oh, no…” Malachi placed his fork on his plate with a clink. Isaiah chuckled.

“Yes, you old grump! If everyone has finished, please follow me to the study!”

His enthusiasm was so infectious all we could do was stand and follow him. Even Malachi finally rose from his chair and joined our little group. We made our way to the study, and Isaiah opened the door with a flourish.

“Twister?” I asked, confused.

“And Mousetrap, Connect Four, Hungry Hungry Hippos, CandyLand…”

Malachi sighed and enlightened us.

“Isaiah has a penchant for 80s board games.”

“Ohhhhh…” We all said, then “Oh.” as realization dawned.

"But… Twister?" I asked again, eyeing the mat with reluctance bordering on dismay. I had always been terrible at board games, especially ones involving coordination.

"Perhaps some brandy in the library first?" Esme offered, tucking her hand in the crook of Isaiah's arm. At his crestfallen expression, she added, sotto voce, "maybe *we'll* play Twister later…" which brightened his face considerably.

"How about Clue?" he tried.

"I'd be more willing to play Clue with a few drinks in me," Kaden said and turned toward the library.

"Deal!" Isaiah agreed and followed with Esme on his arm.

After a couple of drinks, we started with Clue and, as the spirits flowed, our reticence about the other games fell away. By the end of the evening, my cheeks and stomach ached from laughing so hard. Even Malachi seemed to enjoy himself and beat us at Clue twice in a row. We finished the evening watching 007 movies in the home theater and debating the best Bond.

"Okay, it's bedtime." Draco nudged me awake. I had fallen asleep in the theater seat.

“Mm-kay…” I mumbled and let him pull me up. “C’mon, Kaden…” I said, nudging his knee with my foot. He was snoring with his head pitched back, a bowl of popcorn in his lap. He sat up with a snort and nearly knocked the popcorn to the floor, catching it at the last moment.

“Okay! I’m up!” He rubbed his face with his hand and stood.

Draco and I held hands as we all shuffled upstairs to our rooms. We dropped Kaden at his room and made our way to mine. At some point since Draco had been bunking with me, my bed had grown into a king size. The four posters were still there with the High Priestess fresco on the headboard. It was a beautiful bed.

We undressed in silence and got in, snuggling close. I lay on my side, my back pressed against him, and he had an arm tucked around my body.

“Love you more than breathing,” he whispered.

“Mmmm… I love you too, baby.”

~

“Stop here,” Calliope demanded as they drove through a forested area.

Damien scowled briefly but complied.

“What’s going on in that head of yours, witch?”

“I’m going to craft your weapon now.”

Her statement brought a smile to Damien's face.

“I’m... grateful,” he ground out.

“You should be,” she responded and got out. “Stay in the car,” she ordered.

Damien saluted her.

She strode purposefully through the woods. Finding a spot to her liking, she sat and began. Pulling the materials from the ether, she chanted as she constructed. Her hands danced over space and material, crafting a weapon she knew would suit him.

After an hour of work, it began to take form. Her fingers skipped and crafted, and her chanting grew in volume. Blue tendrils of ether swirled and sparked. She smiled when the diaphanous threads began to take the shape of an axe. The belly of the handle was the first to become corporeal. From there, it firmed like a wave in both directions until a wicked, double-bladed battleaxe sat in her lap. Powerful war runes decorated the obsidian blades, and the handle was a flawless oak.

She stood and hefted the weapon, feeling its weight and balance. The blade whistled as it sliced through a sapling. It was perfect.

Damien grew bored after a couple of hours, accompanied by nothing but his thoughts. He whistled along with the radio as he waited for his witch. She was crafting a weapon for him. She did that for precious few. That must mean she cared about him, right? He shook his head, disgusted. He was Damien of The Courts. Everyone trembled at his approach. Other demons feared him. Brought low by a female. It was nauseating. And, he had to admit, terrifying. He had a weakness now, and he knew what happened to those with a weakness. Weaknesses were capitalized on, poked, and twisted by others. Weakness meant vulnerability. Defeat.

He was just about to lose his patience when The Crone emerged from the woods. Her lithe form excited Damien, but what drew his attention was the magnificent battleaxe resting on her shoulder. He jumped from the car and met her in the middle of the road. Her sumptuous lips twisted in a self-satisfied smile, and her eyes twinkled as she looked up at him.

Without a word, she hefted the axe from her shoulder and presented it to Damien with both hands. He ran his hand along the curves of the weapon with a grin. It was a thing of terrible beauty. The obsidian shone deeply enough he could see his reflection. The runes for strength, persistence, victory, and war were etched along the blades. He traced them with a finger in admiration.

"I've never met its equal," he said finally.

"You won't," she responded.

He lifted the axe from her hands and tested the feel. It felt like an extension of his arm. He twisted his wrist and swung it in a tight circle.

"It's perfect," she stated.

"It's perfect," he agreed.

Holding the battleaxe in one hand, he grabbed Calliope with the other and ensnared her in a deep, passionate kiss as the sun set over the water.

They would arrive at Isaiah's under cover of night. Traveling by car had been a point of contention. Damien wanted to slip the Veil and arrive as quickly as possible. The

Crone argued they could sneak up and plan their initial attack better if they drove. Damien capitulated, and they drove.

The closer they got to their destination, the more excited Damien became. All the waiting was nearly over. Soon, his thirst for revenge would be satisfied. He was undecided as to whether he wanted to take Angelica out first or save her for last. The Crone suggested they capture her so he could torture her at his leisure. As had been the case more and more lately, he saw the wisdom of her suggestion.

He was impressed by her. He admired her. The more time he spent with her, the more time he wanted to spend with her. The more he wanted her. It was an ache in his gut, a fire that spread through his limbs and made his nerve endings spark.

He glanced over his shoulder at the battleaxe lying across the back seats. It gleamed in the low light.

"It's thirsty," The Crone said.

"The weapon?"

She nodded and held up a finger before running it along the edge. Damien watched as the blood was absorbed into the obsidian. She smiled.

"Fascinating," Damien said in wonder.

"It will feed and become stronger."

"Are you saying the more I kill with this, the stronger it will become?"

She nodded again. Damien let out a breath.

"You have outdone yourself this time, witch."

"Yes," she replied simply.

They drove in silence as the light grew dim and night overtook the day.

~

Esme was nervous. Isaiah had offered her a separate room, but she had assured him she wanted to sleep with him. She hadn't slept with another person, literally or euphemistically, since she and Isaiah parted all those years ago. She wrung her hands as she watched him prepare the bed.

"Are you okay?" Isaiah paused in arranging the blankets.

"Fine, fine," she responded quickly, dropping her hands to her sides.

"Esme," he said softly.

He let the blanket fall to the bed and walked over to her. She resisted the urge to wring her hands again, unable to meet his eyes.

"Esme," he repeated, tipping her chin up to look at him. He smiled gently. "I haven't had anyone in this room until tonight. I built this room for us. But your comfort is the only thing I care about. I'll sleep on the recliner if you prefer. I just want to be near you."

"I don't know why I'm so nervous."

"I'm nervous too." He shrugged and took her hand, chuckling softly.

"What a pair we are—two of the most powerful beings on the planet acting like adolescents playing 'seven minutes in heaven.' Here," he took her other hand and kissed each of her fingertips. "Just get comfortable and we'll watch some silly TV."

"You don't have a TV in here."

"Easily remedied."

"No, that's okay."

Isaiah brushed her hair back from her face and cupped the nape of her neck with his hand. She let out a shaky breath and gave him a shy smile. He leaned forward and rested his forehead against hers. She let go of his hand and wrapped her arms around his waist with a sigh.

“This feels so good,” she said into his chest.

“It does.”

Esme stretched onto her toes and kissed him. At first, he was surprised, but then he opened his mouth, stroking her tongue with his. He held the back of her head, his fingers tangled in her hair.

She made a small, desperate noise and began tugging at his shirt. She deftly undid the buttons and explored his exposed chest and abs with sure hands. She lightly scratched his back with her nails, and he hummed with satisfaction. His hand slid from the back of her neck, running the back of his fingers down her shoulder and arm, lightly brushing the side of her breast through her shirt.

She broke their kiss just long enough to whip her shirt off before pouncing on him, wrapping her legs around his waist. He laughed, holding her aloft before burying his face into the side of her neck. She wove her arms around his neck, caging his head between her elbows. He pivoted to face the bed and carried her, laying her atop the comforter.

He took a step back to undo his belt, and she watched with rapt attention, biting her bottom lip. He undressed and stepped forward again, nudging her knees apart. He stepped between them, running his palms up her legs and resting his fingers at the waist of her jeans. He tickled her waist lightly, and she giggled, batting his hands away before lifting her hips to allow him to remove her jeans.

She lay before him in her bra and panties, and he took a moment to breathe. She was even more extraordinary than he remembered. She reached her arms out, and he crawled onto the bed, resting on top of her, his hands on the sides of her face. He took a long moment to savor the feel of her skin on his, watching her eyes twinkle with joy.

He kissed her again, reaching a hand down to trace his thumb along the inside of her hip, teasing along the seam of her panties. He pulled the fabric to one side and stroked her wet sex with a groan.

“Please, Isaiah,” she whispered against his lips.

He nodded and pushed one finger inside, curling it to press against her bundle of nerves. She trembled and moaned, lifting her knees to hug his waist. The sounds she made nearly undid him.

"Esme," he pleaded, his voice shaky, "it's been a very, very long time. I'm not going to last much longer."

"I need you inside me, Isaiah. Right now."

He let out a breath and placed his hand on her waist. Slowly, he stroked his way in, inch by inch, until they were one once more.

"Deeper."

"Yes."

They barely moved, their senses heightened to the highest levels. Every breath shifted their bodies slightly, setting off a new explosion of pleasure. Esme arched her back slowly, her head thrown back. He moaned and pushed deeper.

"Isaiah…" Esme rolled her head from side to side, breathing quickly, "Make me cum."

"Anything for you."

He stroked faster, and her breasts bounced, her nipples sliding along his chest. She let out a long moan and came, her muscles spasming and pulsing around him, sending him over the edge with a gasp. He felt himself cum inside her, hot jets of his seed filling her.

They lay together for a long time as their breathing returned to normal.

“Again,” she said.

Kaden awoke with a start, his heart racing. He had a vision.

They were on their way.

He had trouble with one shoe, then abandoned the shoes altogether and raced out of his room in bare feet. He nearly collided with Malachi at the corner.

“You sensed it too?” Kaden asked, breathless.

“My brother draws near,” Malachi confirmed.

Kaden nodded and sprinted towards my room.

“We have to alert the house! I’ll get Pepper and Drake; you rouse Esme and Isaiah!”

Malachi didn’t bother with a response. He just continued on his path to Isaiah’s room.

I was in a twilight sleep when Kaden burst through the door.

“They’re coming! It’s go time!” he yelled.

Instantly awake, Draco threw the blankets aside. We scrambled out of bed, dressing quickly. The air was electric with tension. A shiver of fear ran through my body. This night would either end in victory or in death. I glanced at Draco, strapping a knife to his belt and felt such a rush of love I had to blink back tears.

"They would show up at night," Kaden muttered to himself. "Just like the attack on Christmas. Was that the War of Independence…?"

"K?" I said, trying to keep calm.

"Yeah," Kaden answered absently.

"You're freaking out."

"Of course I am. Aren't you?" He shot me a vulnerable look.

"Yes. But we've trained for this. Even Malachi's stopped saying we're going to die. We can do this."

"I sure hope so."

"Me too."

Once Draco and I were ready, we headed to the library to join the rest of the team. Kaden raced back to his room to fetch his shoes. When we opened the library door, we found Isaiah and Esme locked in an embrace. Her arms were around his waist as he held her to him, one hand cradling her head and the other hugging her back. She was resting her head on his chest with her eyes closed and Isaiah was gently rocking her. It was such an intimate moment that I almost turned away. Instead, I cleared my throat. Esme looked up and smiled.

"Good evening, kids."

"Evening," I replied. "Kaden's getting his shoes. Oh, there he is."

Kaden burst through the door. "Where's Malachi?"

"Right here," Malachi said as he strode into the room.

"Okay," Isaiah said, "so, we're all here. Let's focus."

We gathered in a loose circle, too amped to sit. My hand trembled slightly as I slid it into Draco's slightly damp one. I looked around and saw each of us showing the effect of adrenaline; alert posture and bright eyes. No one was smiling, and all eyes were fixed on Isaiah.

"This is what we've been training for," Isaiah said, his voice steady and commanding. "The Crone and Damien are about to arrive. Watch each other's backs and stick together as much as possible. I've prepared a regiment of golems—use them as cover. Esme and I will handle The Crone and Siren. Kaden, you and Malachi will take on Damien. We outnumber them, but they're formidable and not to be underestimated. Our goal is to separate them and divide and conquer."

Isaiah paused, meeting each of our gazes in turn, the gravity of the moment settling over us all.

"You are strong enough to do this. We are strong enough to do this. They are ancient powers, but so are we. Plus, we have a Gatekeeper—a fact they aren't aware of yet. That's to our advantage. They're coming to us, which is also to our advantage. We know what we're up against, while they don't have all the information. We can use that against them. They don't know there are two Ancients here. They don't know we have a Gatekeeper. We are fighting to protect those we love…"

Isaiah shot a tender glance at Esme.

"And they will never understand that. That is our biggest advantage. We are defending our way of life. We are united. We can win."

Isaiah's brief speech was met with nods and a chorus of agreement. Despite the fear coursing through me, I knew we were ready. I had been preparing for this moment since the slime demon's terrifying words in the sewer months ago—and even longer since we first locked Damien up over sixty years ago.

"They are driving, so they will likely approach from the road," Kaden said.

"Damien has the ability to slip the Veil, so a flank is also possible," Malachi added.

Isaiah conjured a blueprint of the estate and pointed as he spoke.

"We'll pair off: Draco and Angelica, myself and Esme, and Kaden and Malachi, in order to cover the points of ingress."

Isaiah walked over to a corner of the library and rummaged around before producing a handful of rings.

"These are shielding rings. They're not invincible, but they will withstand quite a few attacks before their effectiveness is exhausted."

He handed each of us a ring. I slipped mine on, and the gem glowed green.

"It will turn orange when it's at half strength and purple when it's nearly depleted. Once it's purple, it can take two more hits before it's just jewelry."

He took Esme's hand and placed the ring on her left ring finger with a wink. She smiled slightly and shook her head. If he was scared, he wasn't showing it. His calm demeanor had a soothing effect on me to see him so collected. The result of eons of living, I supposed. I felt another wave of gratitude.

"Thank you, Isaiah. And thank you, Esme and Malachi. This would be a massacre if it were just the three of us," I said, gesturing to Draco, Kaden, and myself.

"It would be," Malachi agreed with a small smile.

"Thanks," I responded, rolling my eyes at him. He shrugged.

"Are we sure we want to split up at the beginning? Shouldn't we meet them outside, like a battlefield?" I asked.

"Kaden makes a good point," Draco offered.

"He does," Isaiah agreed. He pondered for a moment before nodding. "We will meet them as a group. I think we should keep Esme and Angelica back to use the element of surprise regarding Esme and to protect Angelica as long as possible."

I saw the wisdom in Isaiah's suggestion, even though it made me feel like a damsel in distress.

"That's a fine idea, love," Esme said, brushing his cheek with a quick kiss. "They have no reason to think I'm here, and the more cards we keep up our collective sleeve, the better."

~

The Demon screamed up the road, and Damien couldn't keep the smile from his face. Finally. Tonight was the night. He pressed the accelerator harder.

"I know you're excited, but healing a broken neck will delay us. Slow down."

Damien scowled but obeyed. He glanced over at Calliope. She looked radiant in the moonlight filtering through the car windows, clad in her leathers with her whip in her lap. She ran her fingers over the barbs at the end of the whip, cutting and healing herself repeatedly. His heart swelled in his chest. He looked over his shoulder at the battleaxe again, his smile growing wider. She must love him. Nothing as beautiful as that weapon could have been crafted by anyone else. He reached over and grabbed her hand, bringing it to his face to kiss it with flourish. It smelled of blood and magic.

The night sky was clear, the stars bright in the cool air. They parked the car and approached the compound on foot. But just a few steps in, they were grabbed from behind.

“Oh, bother,” The Crone said, her voice bored as she severed the arms of the golem trying to restrain her. With a wave of her hand, she destroyed it. Damien manifested his true form and exploded the golem from within.

“There will be more,” she advised, pursing her lips. “Now I remember. Isaiah was expert at golems and using the weather. Expect it to get cold as well.”

“Child’s play,” Damien scoffed.

“They’re here,” Isaiah announced. “They’ve reached the first of the golems.”

A surge of adrenaline jolted through me, and I twisted the ring on my finger. I scanned the room, noting the bead of sweat on Kaden's brow, Draco’s tense lips, the bored indifference of Malachi, then Isaiah and Esme, seeming as if they were almost having fun. They touched each other in small ways, a brush of the hands, a hand on the small of the back. I took a deep breath.

“Let’s make our way outside,” Isaiah suggested, gesturing to Esme. “After you.”

“What a gentleman,” she replied with a smile, running her hand down his arm as she passed.

We filed out and made our way to the foyer. We wore few weapons, just a knife each on our belts and in our pockets for emergency close combat. I wasn’t great with weapons; I preferred long-distance attacks—runes and spells. My close combat was decent, but I was relying heavily on my compatriots. We aimed to keep Damien and The Crone ignorant to the fact of my being a Gatekeeper for as long as possible.

When we gathered outside, Isaiah began a series of spells, his eyes glowing bright blue. Clouds gathered on the horizon, and the air grew chilly. I watched as the clouds darkened and released their precipitation about three hundred yards from where we stood.

“Throw some lightning in there, darling,” Esme suggested.

Isaiah smiled, and the sky flashed with lightning.

“Oh, and here comes the damned snow,” The Crone grumbled.

“So, our arrival has not gone unnoticed,” Damien said, his voice dripping with sarcasm.

“We knew they would be aware of our presence. The fact I’ve been cracking their wards was a clue enough,” she said, shaking the snow from her dark hair as she trudged on. Damien waded through the snow, which had quickly accumulated to nearly knee-deep. The Crone conjured a fire spell to melt the path ahead, and Damien followed, his battleaxe resting on his shoulder, appreciating the view of her ass in the leathers.

Lightning struck nearby, and The Crone rolled her eyes.

“Really, Isaiah?” she muttered.

She stopped and turned to face Damien. “Remember, I’ll focus on Isaiah. You handle anyone else.”

“Fine,” Damien conceded.

Esme and I returned to the house to prepare for our guerrilla attack.

“We’ll do our best to keep them occupied while you open the Veil and get that demon through. I’ll help you with the rictus spell if necessary. We’ll try to end this thing as quickly as possible." Esme smiled at me as she spoke. She had a calming effect; I found myself smiling back.

"I'm so grateful you're here, Esme."

"Don't mention it. I wouldn't leave Isaiah to handle this alone, anyway."

"You two are amazing together."

"We are. It's silly we've been apart as long as we have."

"Why did you stay apart so long?" I asked, hesitating slightly.

"Pride, mostly. Hurt feelings. Ego. All the wrong reasons, to be honest." Esme tossed her hair back. "Isaiah has been trying to get me to this house for decades. I've just been too proud to acquiesce. When one lives forever, time isn't a factor. I always said, 'maybe next year.' As I said before, silly."

An explosion in the distance made us both turn toward the sound.

"That might be another line of golems falling," Esme observed.

"How many did Isaiah create? A hundred?"

"At least. Probably closer to three hundred. They're great as cannon fodder."

"That makes sense."

I let out a breath. "Are you nervous at all?"

"A little. Calliope is a real bitch to handle. I wouldn't want to face her without Isaiah."

"That's how I feel about Damien. I'm so grateful to have my friends with me. He took out an entire wolf pack before Kaden and I could contain him. It was horrifying. He's a real dick."

Esme chuckled.

"No wonder they're infernal mates. They sound perfect for each other."

"Hooray," I said, trying to lighten the mood.

Another explosion, this time closer, made my nerves buzz.

Damien swung his axe back and forth, taking out golems in long sweeps of the weapon. They were annoying as hell. And it seemed the more he destroyed, the more seemed to appear. He would get overwhelmed, and The Crone would disintegrate them with a spell. Damien was actually enjoying himself. The battleaxe felt good in his hands, and the destruction of the creatures was mildly satisfying, even if they were annoying. It helped him warm up and fueled his bloodlust. The Crone was another matter. She just seemed inconvenienced. Her focus was getting to Isaiah and destroying him. Anything which stood in her way was irritating to her.

They had covered another hundred yards or so, taking them a good half hour due to the relentless golems. The snow had reached their thighs and even in Damien's true form, it was slowing him down. But the trees were becoming thicker, signaling they were getting closer.

Isaiah concentrated, while the others watched him.

"He's incredible to observe," Draco said quietly.

"He's good at what he does, for sure," Kaden agreed.

Malachi rolled his eyes.

"One of the most powerful beings in the world and 'he's good at what he does,'" he muttered.

Kaden tensed, and Malachi shook out his arms.

"What's going on?" Draco asked.

"They're getting closer. They'll be here in twenty minutes," Kaden answered.

Malachi cracked his neck and let his true form manifest. Kaden and Draco gaped.

"Man, you are even more gorgeous than the picture suggested," Kaden said, fanning his face.

"Thank you, I guess," Malachi responded flatly.

"Is Damien as ugly as Malachi is beautiful?" Draco asked.

"Yep," Kaden confirmed. "He's horrifying. He can drive normal people insane just by making them look into his eyes. We have some defense, being mages, but I wouldn't want to spend a lot of time looking at his face."

Esme was holding my hands and helping me meditate, helping me feel for the ley lines around us. With my Vision, she was a pulsing being of cool light. Everything about her was comforting. She felt like a mother, sister, and friend all at once. I adjusted my sight and saw our compatriots on the lawn. I inhaled sharply. Beyond the treeline, the enemy was approaching. Damien's energy was black and oily with flashes of army green. Striding next to him was another being of cool light, but the effect was the opposite of Esme's. It was unsettling and terrifying.

“They’re nearly here,” I told Esme, dropping her hands.

“Don’t lose focus,” she said, peering directly into my eyes. “Remember what Isaiah said. We’re ready. We’re strong. We have numbers and home field advantage on our side. They don’t know I’m here. They don’t know you can send his ugly ass back to Hell. We’re defending our lives and way of life. Love. They only have hate and revenge.”

I nodded and took a deep breath. “I’ll give it everything I have.”

“Me too, honey,” she said, patting my cheek.

When Damien and The Crone emerged from the treeline, the golems surrounding our group in a defensive circle took their fighting stances. There were about one hundred fifty left, and the sight was impressive. The Crone simply laughed.

“Aren’t you happy to see me, Isaiah? It’s been so long. This should be fun,” she called across the grass. Damien chuckled beside her.

Damien scanned the group, his eyes landing on Kaden. “Oh, there he is. That means she’s here somewhere too.” He grinned. “Where’s the little witch, mage?” he boomed across the field. Kaden shuddered. Draco's attention was on his axe.

"And my brother! What a surprise!"

"It's time to go home, Damien," Malachi stated firmly.

"I am home, you feathered ass."

Before Damien could take a step, Isaiah threw up a barrier to impede him. The Crone shattered it with a wave of her hand.

"Now, now, Isaiah," she chided, wagging her finger. "That's no way to treat your guests."

Kaden and Draco chanted a powerful piggybacked attack spell, directing it at Damien. Isaiah launched an attack at The Crone to distract her in order for the boys to get the spell off, hitting Damien directly in the chest. He stumbled back a few steps and fell to one knee, coughing.

"Well, that was just rude," The Crone said, unfurling her cat o' nine tails. In her other hand, she conjured an attack spell and shot it at Kaden. Isaiah countered, catching the attack midway and neutralizing it. Damien roared and charged, his battleaxe held high. The Crone followed quickly behind him.

The golems converged on Damien and The Crone. They fell, one by one, either by magic or my force. After ten minutes, only a dozen remained.

Malachi met them halfway, and The Crone tried to catch him with her whip. He grabbed the whip and yanked, causing her to stumble. Damien swung his axe, the blade whistling through the air. Malachi pivoted, the axe catching the ends of the feathers on his right wing. Isaiah engaged The Crone in a volley of spells, lighting up the night sky.

Esme stood and ran outside the moment she saw Siren stalking the boys from behind. With a war cry, she cast a netting spell to capture Siren, but the cat moved at the last moment, and the spell fell short.

"Why, you sneaky little minx," The Crone called to Esme. "What are you doing here?"

"I wouldn't miss it, Calliope," Esme replied, sending another spell at Siren, who dodged it neatly. Siren pounced on Kaden from behind, dragging him to the ground. I could see the gem on his ring flash with each swipe of her claws until I couldn't take it anymore.

"You want me, you ugly bastard?" I stalked toward the middle of the field, lighting my hands with battle magic.

Damien threw his head back and laughed.

"So there you are! What a pleasant surprise!"

I unleashed a barrage of spells at Siren, trying to get her off Kaden. His gem was orange, and he was pinned on his stomach, taking hit after hit from the cat. Meanwhile, Malachi and Damien were battling full force, with Draco casting spells between their blows. Scorch marks marred Damien's chest, and Malachi was bleeding from multiple shallow cuts. Siren finally had enough of my onslaught and retreated from Kaden's back, fur smoking. Kaden stood and squared off with Siren, his face a mask of fury.

He conjured a giant fireball and hurled it at Siren. It landed squarely on her back, eliciting a scream from The Crone. Isaiah seized the moment of distraction to cast a netting spell on Siren. Esme and The Crone continued their intense exchange of blows, and Esme's gem flashed with each impact of the whip. Retreating a few steps, Esme struggled as The Crone advanced, pressing her advantage.

I hit The Crone from the flank, trying to divert her to lose focus on Esme. Isaiah was securing Siren with a lock on the netting spell and The Crone screamed in outrage again. She hurled a spell at Kaden and he dropped, his gemstone black. My heart fell into my stomach.

“No!!” I howled and doubled my efforts on The Crone.

With the cat secured, Isaiah turned his attention to the battle between Esme and The Crone. Esme took another hit from The Crone’s whip, her gemstone flashing purple. Isaiah charged in from the left, tackling The Crone. She gnashed her teeth and clawed at his face.

I turned to face Damien and lost my breath when his axe caught Draco across the stomach. Malachi hit Damien square in the jaw with a powerful right cross to Damien’s jaw. I lifted off the ground to get to Draco as fast as I could, tapping into the ley lines. Malachi landed another solid punch by the time I reached Draco. Enraged and terrified, I split the Veil right next to Damien. Esme, catching my move, cast a rictus spell on Damien just as I did the same. Isaiah joined in with a punch that lifted The Crone hard enough to lift her feet from the ground and added his spell to ours.

“What the fuck is this?” Damien roared.

“Your way home, brother,” Malachi said, delivering one more punch that sent Damien closer to the tear in the Veil.

“Nooooo!!!” Damien howled as we pushed him closer and closer to the fissure.

A spell struck me from behind, causing me to lose focus for a moment. The tear shimmered and began to shrink. Pain seared through my side, forcing me to sink to the ground. Isaiah ensnared The Crone in a net as she screamed obscenities at him. I refocused on splitting the Veil, and together, we pushed Damien further into the rictus spell, his hooves dragging deep marks in the grass.

We got him closer and closer, and The Crone screamed like a banshee. "Shut up," Esme said, punching her in the throat. The Crone coughed and gagged, and the screaming stopped. All we could hear now was Siren howling in her net and Damien growling and protesting at top volume.

"No!!! Not again, you cunt!"

Ignoring his screams, we pushed with renewed effort. The fissure began to suck him in, and I could see the other side with curious demons looking on, some with a mischievous look as if considering the exit. It was now or never. We collectively pushed again, and Damien fell through, cursing my name as he vanished. I sealed the tear and ran to Draco.

"Check Kaden!" I shouted to Malachi.

Reaching Draco, I rolled him over. Blood was everywhere.

“No, no, no…” I repeated, my hand slipping into the pool of cooling blood. He was unconscious, his skin pale. I searched for a pulse and began to cry hysterically when I couldn’t find one.

“We need Mischief!” I cried out in desperation.

“Here she is, honey,” Esme said gently, placing Mischief on Draco’s chest.

“Oh, sweet girl. I really need your help.”

I’m here, mama.

Mischief placed her paws on Draco’s wound, and a blend of hot and cool energy flowed from her as she worked to heal him.

“Oh, God…” I sat back, pressing the heel of my blood-covered hand against my forehead.

“We need her over here, too,” Malachi called from where Kaden lay.

My heart broke.

“No, not both of my guys…”

The night took on a surreal quality, and a strange calm settled over me. I felt numb. This couldn't be real. Malachi's hand on my shoulder was the only tangible connection to reality.

"I'll do what I can for Kaden, but healing isn't my strength," he said. "Esme!"

Esme jogged over, her face etched with concern. "How are your healing skills?" Malachi asked her.

"Not great, unfortunately," Esme frowned.

Isaiah was still by The Crone and Siren, making sure they were secure. He strode over to Kaden. He placed a hand on Kaden's chest, and his shoulders fell in defeat.

"No!!!" I keened. "Damn him! He would have seen this coming! Why did he fight??"

The Crone started cackling from her prison.

"Yet she lost everything anyway!" She sing-songed. Then hummed a happy tune to herself.

Mama… Mischief's voice sounded small and sad.

"No, baby. Keep trying," I urged her.

She nodded and focused all her effort on healing Draco.

Hey, Pepper ~

By the time you're reading this, you'll know I didn't survive the battle. I know you're pissed at me, but I saw the whole battle, and my death was the only way we were going to win. Without me, The Crone gets a hold of you and basically plays hot potato with you with Damien. I've seen all the outcomes. This was the only way.

You are my best friend, and I love you with all my soul. You've been the best friend a foundling like me could ever have. Please don't let this break your heart.

-K

When I got Draco to bed, I found a letter on my pillow. I didn't know when he had a chance to place it there, but Kaden had written me an apology/suicide note. My heart tore in two as I read it. My best friend. I couldn't imagine my life without him. He sacrificed himself to protect me. Just like him. He always had my back, and he rescued me one last time. Draco held my hand as tears streamed down my face.

That night, Mischief wobbled in, climbed onto the bed, and passed out on a pillow. She slept until she needed food or water, then fell back asleep. She deserved it. She had saved my love. If I had lost both of my guys, I'm not sure I'd survive. Losing Kaden was devastating enough.

Malachi left without saying 'goodbye.' Just flew his way back to defending humanity or whatever he did. It was low on my list of sad things these days.

Isaiah offered us a place as long as we needed, but after a month or so, it was time to return to our lives. Whatever that looked like without my best friend, Kaden. I was moving back home with Draco and Mischief to Hollywood—a small joy in the midst of the grief. My first thought was to tell Kaden.

We stayed to witness Isaiah and Esme's binding ceremony. It was beautiful. Kaden would have loved it.

"We're home, sweetheart."

Draco winced as he put the car into park. He was still sore from being nearly disemboweled by Damien and his giant axe. Mischief had done all she could, but his healing had to be done by his body. I sat for a moment, Mischief on my lap, and sighed.

I kissed her forehead, feeling a bittersweet comfort. I was happy to be home.

Made in the USA
Monee, IL
04 November 2026

67387580R00266